Sintering

To Youngshin, Joohyun, and Taeho

Sintering

Densification, Grain Growth, and Microstructure

Suk-Joong L. Kang

ELSEVIER
BUTTERWORTH
HEINEMANN

AMSTERDAM • BOSTON • HEIDELBERG • LONDON • NEW YORK • OXFORD
PARIS • SAN DIEGO • SAN FRANCISCO • SINGAPORE • SYDNEY • TOKYO

Elsevier Butterworth-Heinemann
Linacre House, Jordan Hill, Oxford OX2 8DP
30 Corporate Drive, Burlington, MA 01803

First published 2005

British Library Cataloguing in Publication Data
A catalogue record for this book is available from the British Library

Library of Congress Cataloging in Publication Data
A catalogue record for this book is available from the Library of Congress

ISBN 07506 63855

For information on all Elsevier Butterworth-Heinemann
publications visit our website at http://books.elsevier.com

Typeset by Keyword Group
Transferred to Digital Printing 2008

Resources to accompany this book

Worked solutions to selected problems in this book are available
online for tutors and lecturers who either adopt or recommend
the text. Please visit http://books.elsevier.com/manuals and follow
the registration and login instructions on-screen.

Working together to grow
libraries in developing countries

www.elsevier.com | www.bookaid.com | www.sabre.com

ELSEVIER   BOOK AID
            International   Sabre Foundation

CONTENTS

v

PREFACE

Sintering is a technique of consolidating powder compacts by use of thermal energy. This technique is one of the oldest human technologies, originating in the prehistoric era with the firing of pottery. These days, sintering is widely used to fabricate bulk ceramic components and powder metallurgical parts.

To understand sintering and solve related problems, a fundamental and comprehensive understanding of the basic principles of materials science is needed. During the teaching of a course on sintering to graduate students at KAIST (Korea Advanced Institute of Science and Technology) for many years, the author has recognized that few books are available that correlate sintering with the basic principles of materials science. In many cases, being considered to be a technique of material processing, sintering is briefly described in a chapter in books on ceramic processing or powder metallurgy. Otherwise, the books dedicated solely to sintering are often too bulky and present a number of experimental data with little explanation of the general principles. These books offer a guide to researchers and engineers rather than to students.

Based on the author's lecture notes on sintering, which had been offered to KAIST students for more than 12 years, a Korean book was written in 1997. The book was concerned with the fundamentals of sintering phenomena including densification, grain growth and related microstructures. The present book is an enlarged, English version of the Korean book. The latest research results on sintering, in particular on grain growth, and recent literatures are included in the English version. This book is written primarily as a textbook for graduate students and as a reference work for researchers. However, the book can also be used as a reference book for a ceramic processing or powder metallurgy course for senior or junior students, as in many Korean Universities.

This book aims to provide readers with basic principles of densification and grain growth, and eventually of microstructure and its development during sintering. This book consists of six Parts and describes both densification and grain growth with the same emphasis, unlike most other books where grain growth is treated as a minor subject. Fundamentals of densification and grain growth are treated separately, and then both together. Related microstructural development is also described. This scheme is maintained throughout the book both for solid state sintering (Parts II, III and IV) and for liquid phase sintering (Part VI). This is one of the reasons why the subtitle of this book is *Densification, Grain Growth and Microstructure*. The basis of sintering science, including thermodynamics of interfaces and microstructure of polycrystals, is presented at the beginning of the book (Part I) to provide the readers with basic knowledge for the study of sintering. Part V is dedicated to densification of ionic crystals. At the end of each Part not only theoretical but also practical problems are included to assist readers in understanding sintering phenomena that occur in real systems. The nature of the problems is also related to the application of the principles and therefore mostly explanatory rather than calculative.

In writing this book I have been indebted to a number of people. Without their support, help and encouragement this book could not have been realized. I acknowledge Professor Duk Yong Yoon, as my teacher and colleague, for introducing sintering science to me and for studying sintering and micro-structural problems together. I also wish to acknowledge Professor Richard J. Brook for inspiring and encouraging me in studying sintering science. A part of this book was written when I was on leave at the University of New South Wales, Sydney. The Australian Research Council and my host, Professor Janusz Nowotny, are greatly acknowledged for their support. Writing in English has been a hard job for me. I am very grateful to Professor Emeritus Max Hatherly and to Dr John G. Fisher for reading the whole manuscript and improving the text. If the writing is below standard, it is solely due to my originally poor writing. Professor Jürgen Rödel is acknowledged for his contributions to the 'Constrained Sintering' section. My colleagues, Professors Doh-Yeon Kim, Han-Ill Yoo, Nong-Moon Hwang and Ho-Yong Lee, are also acknowledged for helpful discussions. Most of the figures in the book were prepared and redrawn by Dr Young-Woo Rhee, my former student, and Mr Yang-Il Jung to whom I should like to express my sincere thanks. I also wish to thank Ms Eun-Ju Kim for her typing assistance. Last, but not least, I would like to thank my former and present students who have studied sintering together with me over the past 20 years.

Suk-Joong L. Kang
Daejeon
April 2004

PART I
BASIS OF SINTERING SCIENCE

When thermal energy is applied to a powder compact, the compact is densified and the average grain size increases. The basic phenomena occurring during this process, called sintering, are densification and grain growth. To understand sintering and utilize it in materials processing, we first need to understand the fundamentals of the thermodynamics and kinetics of the two basic phenomena. In Part I, the process of sintering is briefly described and the related thermodynamics presented. The polycrystalline microstructure obtained after sintering is also characterized. The kinetics which determines the development of microstructure during sintering is the main content of this book and will be described extensively in Parts II to VI.

PART I
BASIS OF SINTERING SCIENCE

I

SINTERING PROCESSES

I.I WHAT IS SINTERING?

Sintering is a processing technique used to produce density-controlled materials and components from metal or/and ceramic powders by applying thermal energy. Hence, sintering is categorized in the synthesis/processing element among the four basic elements of materials science and engineering, as shown in Figure 1.1.[1] As material synthesis and processing have become crucial in recent years for materials development, the importance of sintering is increasing as a material processing technology.

Sintering is, in fact, one of the oldest human technologies, originating in the prehistoric era with the firing of pottery. The production of tools from sponge iron was also made possible by sintering. Nevertheless, it was only after the 1940s that sintering was studied fundamentally and scientifically. Since then, remarkable developments in sintering science have been made. One of the most important and beneficial uses of sintering in the modern era is the fabrication of sintered parts of all kinds, including powder-metallurgical parts and bulk ceramic components.

Figure 1.2 shows the general fabrication pattern of sintered parts. Unlike other processing technologies, various processing steps and variables need to be considered for the production of such parts. For example, in the shaping step, one may use simple die compaction, isostatic pressing, slip casting, injection moulding, etc., according to the shape and properties required for the end product. Depending on the shaping techniques used, not only the sintering conditions but also the sintered properties may vary considerably. In the sintering step, too, there are various techniques and processing variables; variations in sintered microstructure and properties can result.

Sintering aims, in general, to produce sintered parts with reproducible and, if possible, designed microstructure through control of sintering variables. Microstructural control means the control of grain size, sintered density, and

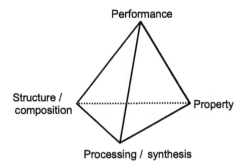

Figure I.I. The four basic elements of materials science and engineering.[1]

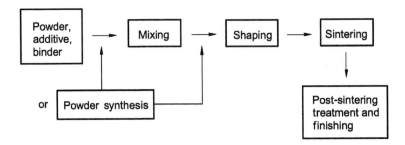

Figure I.2. General fabrication pattern of sintered parts.

size and distribution of other phases including pores. In most cases, the final goal of microstructural control is to prepare a fully dense body with a fine grain structure.

I.2 CATEGORIES OF SINTERING

Basically, sintering processes can be divided into two types: solid state sintering and liquid phase sintering. Solid state sintering occurs when the powder compact is densified wholly in a solid state at the sintering temperature, while liquid phase sintering occurs when a liquid phase is present in the powder compact during sintering. Figure 1.3 illustrates the two cases in a schematic phase diagram.* At temperature T_1, solid state sintering occurs in an A–B powder compact with composition X_1, while at temperature T_3, liquid phase sintering occurs in the same powder compact.

In addition to solid state and liquid phase sintering, other types of sintering, for example, transient liquid phase sintering and viscous flow sintering, can be

---

*Here, various types of sintering are explained using a schematic phase diagram, although, in most cases, the proper sintering type depends on the material system and/or the sintering purpose.

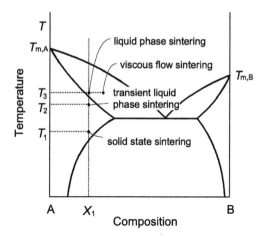

Figure 1.3. Illustration of various types of sintering.

utilized. Viscous flow sintering occurs when the volume fraction of liquid is sufficiently high, so that the full densification of the compact can be achieved by a viscous flow of grain–liquid mixture without having any grain shape change during densification. Transient liquid phase sintering is a combination of liquid phase sintering and solid state sintering. In this sintering technique a liquid phase forms in the compact at an early stage of sintering, but the liquid disappears as sintering proceeds and densification is completed in the solid state. An example of transient liquid phase sintering may also be found in the schematic phase diagram in Figure 1.3 when an A–B powder compact with composition X_1 is sintered above the eutectic temperature but below a solidus line, for example at temperature T_2. Since the sintering temperature is above the A–B eutectic temperature, a liquid phase is formed through a reaction between the A and B powders during heating of the compact. During sintering, however, the liquid phase disappears and only a solid phase remains because the equilibrium phase under the given sintering condition is a solid phase.

In general, compared with solid state sintering, liquid phase sintering allows easy control of microstructure and reduction in processing cost, but degrades some important properties, for example, mechanical properties. In contrast, many specific products utilize properties of the grain boundary phase and, hence, need to be sintered in the presence of a liquid phase. Zinc oxide varistors and $SrTiO_3$ based boundary layer capacitors are two examples. In these cases, the composition and amount of liquid phase are of prime importance in controlling the sintered microstructure and properties.

Figure 1.4 shows typical microstructures of partially sintered powder compacts without (a) and with (b) a liquid phase. In both cases, sintering has proceeded to the final stage in which pores are isolated. Such an isolated pore stage is generally reached quickly at usual sintering temperatures.

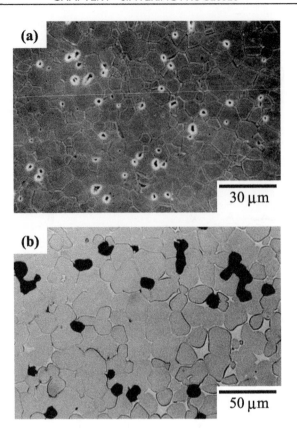

Figure I.4. Typical microstructures observed during (a) solid state sintering (Al_2O_3) and (b) liquid phase sintering (98W-1Ni-1Fe(wt%)).

The elimination of isolated pores is more time consuming and utilizes almost all of the sintering time.

I.3 DRIVING FORCE AND BASIC PHENOMENA

The driving force of sintering is the reduction of the total interfacial energy. The total interfacial energy of a powder compact is expressed as γA, where γ is the specific surface (interface) energy and A the total surface (interface) area of the compact. The reduction of the total energy can be expressed as

$$\Delta(\gamma A) = \Delta\gamma A + \gamma \Delta A \qquad (1.1)$$

Here, the change in interfacial energy ($\Delta\gamma$) is due to densification and the change in interfacial area is due to grain coarsening. For solid state sintering, $\Delta\gamma$ is related to the replacement of solid/vapour interfaces (surface) by

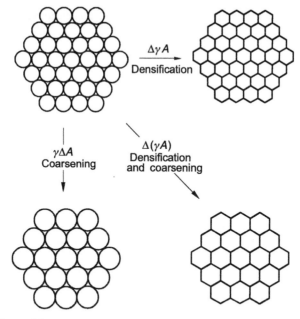

Figure I.5. Basic phenomena occurring during sintering under the driving force for sintering, $\Delta(\gamma A)$.

solid/solid interfaces. As schematically shown in Figure 1.5, the reduction in total interfacial energy occurs via densification and grain growth, the basic phenomena of sintering.

In general, the size of powders for sintering is in the range between 0.1 and 100 μm; the total surface energy of the powder is 500–0.5 J/mole. This energy is inconsiderably small, compared with the energy change in oxide formation which is usually in the range between 300 and 1500 kJ/mole. If the desired microstructure of the sintered body is to be achieved by the use of such a very small amount of energy, it is necessary to understand and control the variables involved in the sintering processes.

I.4 SINTERING VARIABLES

The major variables which determine sinterability and the sintered microstructure of a powder compact may be divided into two categories: material variables and process variables (Table 1.1). The variables related to raw materials (material variables) include chemical composition of powder compact, powder size, powder shape, powder size distribution, degree of powder agglomeration, etc. These variables influence the powder compressibility and sinterability (densification and grain growth). In particular, for compacts containing more than two kinds of powders, the homogeneity of the

Table I.I. Variables affecting sinterability and microstructure

Variables related to raw materials (material variables)	Powder:
	  shape, size, size distribution, agglomeration, mixedness, etc.
	Chemistry:
	  composition, impurity, non-stoichiometry, homogeneity, etc.

Variables related to sintering condition (process variables)	Temperature, time, pressure, atmosphere, heating and cooling rate, etc.

powder mixture is of prime importance. To improve the homogeneity, not only mechanical milling but also chemical processing, such as sol-gel and coprecipitation processes, have been investigated and utilized. The other variables involved in sintering are mostly thermodynamic variables, such as temperature, time, atmosphere, pressure, heating and cooling rate. Many previous sintering studies have examined the effects of sintering temperature and time on sinterability of powder compacts. It appears, however, that in real processing, the effects of sintering atmosphere and pressure are much more complicated and important. Unconventional processes controlling these variables have also been intensively studied and developed (see Section 5.6 and Section 11.6).

2

THERMODYNAMICS OF THE INTERFACE[2]

2.1 SURFACE ENERGY AND ADSORPTION

2.1.1 Surface Energy

In what follows 'surface' is defined as the plane between condensed matter and a vapour phase or vacuum, such as solid/vapour and liquid/vapour interfaces. In a broader sense, the term 'interface' is used for the dividing plane between any two different phases. The existence of an interface means, by itself, the presence of an excess interface energy over the bulk energy. Since the driving force for sintering is the reduction of the total interfacial energy of the system concerned, it will be useful to understand the thermodynamic characteristics of interfacial energy.

Figure 2.1(a) is a schematic of a system containing an interface. Here, σ is the interface and α and β homogeneous phases. The interface in a real system can be represented thermodynamically as a dividing surface σ, although the variation in chemical composition at the interface occurs across multiatomic layers, as shown schematically in Figure 2.1(b). For the system in Figure 2.1(a), any extensive thermodynamic property Φ can then be expressed as the sum of the properties of the bulk phases and that of the interface.

$$\Phi = \Phi^\alpha + \Phi^\beta + \Phi^\sigma \tag{2.1}$$

Here, Φ^σ may be regarded as an additional term due to the presence of a transition layer which is determined by the curvature and area of the interface.

Hence, an infinitesimal change in the internal energy at the surface, which is an excess surface quantity, can be expressed as

$$dE^\sigma = TdS^\sigma + \sum_{i=1}^{m} \mu_i dn_i^\sigma + \gamma dA \tag{2.2}$$

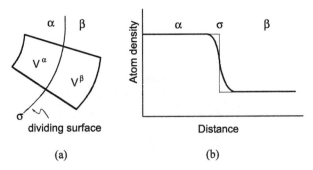

Figure 2.I. Schematic of (a) a system containing a phase α, a phase β and an interface σ, and (b) atom density near the interface.

if we choose a dividing surface where the effect of curvature is negligible. Here, S is the entropy, μ_i the chemical potential of component i and n_i the number of moles of component i. It can be seen from Eq. (2.2) that surface energy γ is the partial derivative of surface internal energy E^σ with respect to surface area A. The surface energy can also be expressed as the change in the total internal energy of the system.

$$dE = dE^\alpha + dE^\beta + dE^\sigma$$

$$= TdS + \sum_{i=1}^{m} \mu_i dn_i - P^\alpha dV^\alpha - P^\beta dV^\beta + \gamma dA \tag{2.3}$$

where P is the pressure and V the volume. Hence,

$$\gamma \equiv \left(\frac{\partial E}{\partial A}\right)_{S,n_i,V^\alpha,V^\beta} \tag{2.4}$$

Equation (2.4), the thermodynamic definition of γ, shows that γ is the reversible work required to create a unit area of the surface.

The internal energy at surface E^σ is expressed as

$$E^\sigma = TS^\sigma + \sum_{i=1}^{m} n_i^\sigma \mu_i + \gamma A \tag{2.5}$$

From this equation and Eq. (2.2),

$$d\gamma = -\frac{S^\alpha}{A}dT - \sum_{i=1}^{m} \frac{n_i^\sigma}{A}d\mu_i$$

$$\equiv -\frac{S^\sigma}{A}dT - \sum_{i=1}^{m} \Gamma_i d\mu_i \tag{2.6}$$

where Γ_i is the number of excess moles of i adsorbed at a unit surface area. Equation (2.6) is known as the Gibbs adsorption equation.

2.1.2 Surface Energy and Thermodynamic Potential

The thermodynamic potential Ω is defined as

$$\Omega \equiv F - \sum_{i=1}^{m} n_i \mu_i \tag{2.7}$$

where F is the Helmholtz free energy, and represents the reversible work in a system at constant temperature, volume and chemical potential. Since

$$\sum_{i=1}^{m} n_i \mu_i = G \tag{2.8}$$

for bulk homogeneous phases, $\Omega^\alpha = -P^\alpha V^\alpha$ and $\Omega^\beta = -P^\beta V^\beta$. Then, the surface excess potential Ω^σ is expressed as

$$\Omega^\sigma = \Omega - \Omega^\alpha - \Omega^\beta = F + P^\alpha V^\alpha + P^\beta V^\beta - \sum_{i=1}^{m} n_i \mu_i$$

$$= \gamma A = F^\sigma - \sum_{i=1}^{m} n_i^\sigma \mu_i \tag{2.9}$$

Consequently,

$$\gamma = \frac{F^\sigma}{A} - \sum_{i=1}^{m} \Gamma_i \mu_i \equiv f^\sigma - \sum_{i=1}^{m} \Gamma_i \mu_i \tag{2.10}$$

This equation indicates that the surface energy is the work required to create unit surface area under constant temperature, volume and chemical potential.

2.1.3 Relative Adsorption

The relative adsorption of i with respect to component 1, $\Gamma_i^{(1)}$, is defined as

$$\Gamma_i^{(1)} \equiv \Gamma_i - \Gamma_1 \frac{C_i^\alpha - C_i^\beta}{C_1^\alpha - C_1^\beta} \tag{2.11}$$

where C_i^α and C_i^β are the numbers of moles of component i in a unit volume of α and β phases, respectively. In this equation, $\Gamma_i^{(1)}$ is independent of the position of the dividing surface and this surface may then be taken as the plane where Γ_1 is zero. Hence, the Gibbs relative adsorption $\Gamma_i^{(1)}$ may be considered as the adsorption of i at the surface where the adsorption of 1 is zero. In this regard, $\gamma = f^{\alpha}$ for a single component system while γ is affected by the adsorption of all components except component 1 for a multicomponent system. For the simple case of monolayer adsorption at a solid/vapour or liquid/vapour interface, $\Gamma_i^{(1)}$ is simplified as

$$\Gamma_i^{(1)} = \frac{n^{(m)}}{A}\left(X_i^{(m)} - X_1^{(m)}\frac{X_i^{\alpha}}{X_1^{\alpha}}\right) \tag{2.12}$$

where X is the mole fraction in the monolayer (m).

2.2 SURFACE TENSION AND SURFACE ENERGY

The existence of attractive forces between atoms at a surface (surface tension force*) may be visualized in an experiment using a liquid film. The surface tension of a liquid film is the force per unit length of a wire, σ_{xx}, used to extend the film area, as shown in Figure 2.2. When the length of the liquid film is increased by dx, the work done by the extension, W, is

$$W = 2\sigma_{xx}l\mathrm{d}x \tag{2.13}$$

where l is the length of the wire. Here, the atom density at the film surface is unchanged with the extension because the surface atom density is constant. This results from the fact that, during the extension of a liquid film, the supply of atoms from bulk to the surface is easy thanks to the high mobility of liquid atoms. Since the density of surface atoms is constant, irrespective of liquid film area, the increase in total surface energy, ΔE, due to surface area increase is expressed as

$$\Delta E = \Delta A \gamma = 2\gamma l\mathrm{d}x \tag{2.14}$$

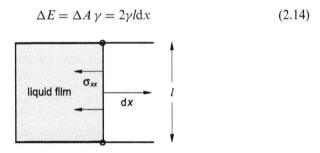

Figure 2.2. Schematic for the calculation of the correlation between surface tension and surface energy in a fluid.

---

*The term 'surface tension' is often used as the synonym of 'surface energy'. In this book, however, it will be used as the force between atoms parallel to the surface. Therefore, the surface tension is interpreted as the force per unit length and may be regarded as the surface stress.

Therefore, for liquids, the surface tension (N/m) has the same value as the surface energy (J/m^2), although they are basically different physical properties. If the film is extended in the y-direction, the exerted surface tension is σ_{yy} which has the same value as σ_{xx}. In general, the surface tension σ of any surface is taken as the average of σ_{xx} and σ_{yy}.

The surface tension of a liquid is due to the fact that the interatomic distance is greater at the surface than in the bulk. It follows that, at the surface itself, a negative pressure parallel to the surface is exerted and surface tension is determined, therefore, by the atom arrangement at the surface. There is no stress perpendicular to the surface and the liquid surface is in a state of plane stress.

For liquids, increase in surface area is achieved by the fast supply of atoms from the bulk to the surface and the atom arrangement with surface area increase is invariable and isotropic as the increase occurs. For solids, on the other hand, surface area increase is achievable by atom supply from the bulk and also by an increase in interatomic distance at the surface. In other words, the atom arrangement at the surface of a solid and, hence, the surface tension can vary with surface area change.

Therefore, surface tension takes the same value as surface energy only when the atom arrangement at the surface is independent of external stress in a reasonable period of time and this conclusion is satisfied when the atom mobility is high enough. The relationship between surface tension and surface energy is determined by the atomic arrangement at the surface and this can be understood by introducing macroscopic strains to a body.

On the basis that the energy state of a unit cube after being cut into two followed by a reversible deformation is the same as that of a unit cube after a reversible deformation followed by being cut into two, Mullins[3] derived an equation which relates surface tension with surface energy as

$$\sigma_{ij} = \delta_{ij}\gamma + \frac{\partial\gamma}{\partial\varepsilon_{ij}} \tag{2.15}$$

where $\partial\varepsilon_{ij}$ is the amount of strain per unit length and δ_{ij} is the Kronecker delta ($\delta_{ij}=1$ for $i=j$ and $\delta_{ij}=0$ for $i\neq j$). In the partial derivative in Eq. (2.15), all other deformations except ε_{ij} are zero. Even for solids, if the temperature is high enough to satisfy $d\gamma/d\varepsilon_{ij}=0$ in a reasonable time span, the surface energy can be measured by, for example, the zero-creep technique.[4] Since powder compacts are usually sintered above two-thirds of the homologous temperature ($T_H = T/T_m$), surface tension can be considered to be the same as surface energy in conventional sintering processes. Although the surface energy varies with crystallographic orientation, it is more or less 1 J/m^2 for most metals and ceramics.[4-6]

2.3 THERMODYNAMICS OF CURVED INTERFACES

2.3.1 Capillarity and Atom Activity

Consider a system where two phases, α and β, are separated by a curved interface and are in equilibrium. If the total volume V, temperature T and chemical potential μ_i are constant in this system, the change in thermodynamic potential caused by an infinitesimal movement of the interface is null. That is

$$d\Omega = 0 = d\Omega^\alpha + d\Omega^\beta + d\Omega^\sigma$$

$$= -P^\alpha dV^\alpha - P^\beta dV^\beta + \gamma dA \tag{2.16}$$

Since $dV = 0$,

$$P^\alpha - P^\beta = \gamma \frac{dA}{dV^\alpha}$$

$$= \gamma K \tag{2.17}$$

Here, K is the average curvature of the interface. Let r_1 and r_2 be the radii of curvature perpendicular with each other at the interface,

$$K = \left(\frac{1}{r_1} + \frac{1}{r_2}\right) \tag{2.18}$$

Therefore, in general,

$$P^\alpha - P^\beta = \left(\frac{1}{r_1} + \frac{1}{r_2}\right)\gamma \tag{2.19}$$

This equation is the well-known LaPlace equation (also referred to as the Young–LaPlace equation). For a sphere, Eq. (2.19) becomes

$$P^\alpha - P^\beta = \frac{2}{r}\gamma \tag{2.20}$$

This equation can also be derived easily by considering the fact that the work required to infinitesimally expand an air bubble in water is stored as an increase in total interfacial energy provided that the process occurs under equilibrium.[5] Equation (2.20) applies for all phase components. For example, the pressure in a water drop relative to its surrounding pressure is the same as the pressure in a gas bubble in water relative to the water pressure, provided only that the sizes of the water drop and the gas bubble are the same.

For a single component system, when α and β phases join together with a curved interface and are in equilibrium, the equation

$$\mu^{\alpha}(T, P^{\alpha}) = \mu^{\alpha}\left(T, P^{\beta} + \gamma K\right)$$
$$= \mu^{\beta}\left(T, P^{\beta}\right) \tag{2.21}$$

holds. For an incompressible α phase, a Taylor series expansion of $\mu^{\alpha}(T, P^{\beta} + \gamma K)$ results in

$$\mu^{\alpha}\left(T, P^{\beta} + \gamma K\right) = \mu^{\alpha}\left(T, P^{\beta}\right) + \gamma K V_m^{\alpha} \tag{2.22}$$

Here, V_m^{α} is the molar volume of α. Therefore, from Eq. (2.21),

$$\mu^{\alpha}\left(T, P^{\beta}\right) - \mu^{\beta}\left(T, P^{\beta}\right) + \gamma K V_m^{\alpha} = 0 \tag{2.23}$$

Equation (2.23) shows, in fact, the relation between the bulk chemical potentials of α and β, and the interfacial energy. Equation (2.22) can also be expressed as

$$\mu_r^{\alpha} = \mu_{\infty}^{\alpha} + \gamma K V_m^{\alpha} \tag{2.24a}$$

Then,

$$\mu_r^{\beta} = \mu_{\infty}^{\beta} \tag{2.24b}$$

Equation (2.24a) is referred to as the Gibbs–Thompson equation or Thompson–Freundlich equation. This equation is correct only when the effect of the interface is exerted solely on the α phase, i.e. the interface belongs only to the α phase. Therefore, when using Eq. (2.24a), Eq. (2.24b) must be simultaneously satisfied. When we express Eq. (2.24a) in terms of atom activity a,

$$RT \ln \frac{a_r}{a_{\infty}} = \gamma K V_m^{\alpha} \tag{2.25a}$$

If $(a_r - a_{\infty})/a_{\infty} \ll 1$

$$a_r \cong a_{\infty}\left(1 + \frac{2\gamma V_m^{\alpha}}{RTr}\right) \tag{2.25b}$$

2.3.2 Application to Sintering: Condensed and Dispersed Phases

For a single component system, since $d\mu^\alpha = d\mu^\beta$ for an infinitesimal and reversible change between α and β phases which are in equilibrium at a given temperature,

$$V_m^\alpha dP^\alpha = V_m^\beta dP^\beta \tag{2.26}$$

and

$$V_m^\alpha d(P^\alpha - P^\beta) - (V_m^\beta - V_m^\alpha)dP^\beta = 0 \tag{2.27}$$

Since $P^\alpha - P^\beta = \gamma K$, the integration of Eq. (2.27) from 0 to K and from P_0 to P^β gives

$$P^\beta - P_0 = \gamma K \frac{V_m^\alpha}{V_m^\beta - V_m^\alpha} \tag{2.28a}$$

Similarly,

$$P^\alpha - P_0 = \gamma K \frac{V_m^\beta}{V_m^\beta - V_m^\alpha} \tag{2.28b}$$

When α is a condensed phase such as a solid or liquid and β is a dispersed phase that follows the ideal gas law,

$$P^\alpha = P_\infty + \frac{2\gamma}{r} \tag{2.29a}$$

and

$$p^\beta = p_\infty + \frac{2\gamma}{r}\frac{p_\infty V_m^\alpha}{RT} = p_\infty\left(1 + \frac{2\gamma V_m^\alpha}{RTr}\right) \tag{2.29b}$$

because $V_m^\alpha \ll V_m^\beta$. Here p^β is the vapour pressure of α.

Equations (2.29a) and (2.29b) are probably the most important and most basic equations for the explanation of sintering phenomena. Figure 2.3 illustrates how Eq. (2.29) is applied to sintering. If a local equilibrium between a condensed and a dispersed phase is maintained across a curved interface (Figure 2.3),* the pressure in region I with a positive curvature is higher than

---

*This assumption, which is a basic assumption to explain sintering phenomena, implies that material transport determines the overall kinetics of any change. It is also a general assumption in treating diffusion-controlled kinetics in materials science.

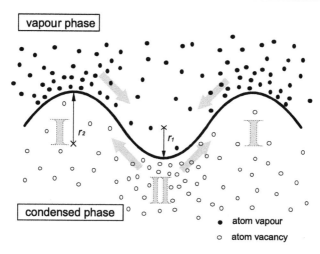

Figure 2.3. Schematic showing distributions of vacancies (○) and vapour atoms (●) near a curved interface.

that in region II with a negative curvature according to Eq. (2.29a). In addition, because of the pressure difference, the vapour pressure above region I is higher than that above region II according to Eq. (2.29b). On the other hand, for a vacancy, which can be considered as a dispersed phase of a vacuum, region I has a negative curvature and region II a positive curvature from the vacuum side. The equilibrium vacancy concentration in region II is higher than that in region I. The concentration difference can be calculated from Eq. (2.29b). Therefore, Eq. (2.29) can be used to calculate the difference in not only the vapour pressure of materials but also the vacancy concentration in the bulk.

The thermodynamic driving force for sintering is the reduction in total interfacial energy. However, in terms of kinetics, the differences in bulk pressure, vapour pressure and vacancy concentration due to interface curvature induce material transport. Note that the three kinds of thermodynamic phenomena (differences in bulk pressure, vapour pressure and vacancy concentration) occur simultaneously and independently.

2.3.3 Energy Change Due to a Curved Interface

So far, we have assumed no energy change in the condensed phase due to capillary pressure. In other words, the condensed phase is assumed to be incompressible and its volume does not change with capillary pressure. In real systems, however, the condensed phase is also compressible and its energy increases with an increased capillary pressure. The energy increase by deformation of a condensed phase of 1 mole is

$$W = -\int_0^P P\,\mathrm{d}V = -\int_0^P P\left(\frac{\partial V_m}{\partial P}\right)_T \mathrm{d}P = V_m\kappa \int_0^P P\,\mathrm{d}P$$

$$= \frac{1}{2}V_m\kappa P^2 \tag{2.30}$$

where V_m is the molar volume and κ the compressibility $\left(= -(\partial V_m/\partial P)_T/V_m\right)$. Compared with the total surface energy, the deformation energy calculated by Eq. (2.30) is negligible for powders of larger than nanometer size. It follows, therefore, that the energy difference between a powder compact and its sintered bulk is the difference in total interfacial energy between the two bodies.

3

POLYCRYSTALLINE MICROSTRUCTURES

3.1 INTERFACIAL TENSION AND MICROSTRUCTURE

The polycrystalline microstructure is the microstructure we observe after sintering a powder compact to full density. In homogeneous polycrystalline materials, this microstructure is determined, in general, by the interfacial tension.* Figure 3.1 illustrates an equilibrium state between three interfacial tensions. For this geometry, the sine law is satisfied so that

$$\frac{\sigma_{12}}{\sin \theta_3} = \frac{\sigma_{23}}{\sin \theta_1} = \frac{\sigma_{31}}{\sin \theta_2} \tag{3.1}$$

where σ_{ij} is the tension force per unit length acting on a phase k.

3.1.1 Wetting Angle

The wetting angle is defined as the angle between solid/liquid and liquid/ vapour interfaces when a liquid drop is placed on a solid substrate, as shown in Figure 3.2. When we consider a force equilibrium in the direction parallel to the substrate,

$$\gamma_s = \gamma_{sl} + \gamma_l \cos \theta \tag{3.2}$$

is satisfied if we assume that $\sigma = \gamma$. Here, θ is the wetting angle, γ_s the solid surface energy, γ_l the liquid surface energy and γ_{sl} the solid/liquid interfacial energy. In Figure 3.2, since a force balance is not satisfied in the vertical

---

*In materials with high anisotropy in the grain boundary energy and, thus, with faceted grain boundaries, the angles between boundaries may not be uniquely defined by an interfacial tension condition because of torque on facets.[7,8]

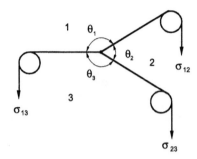

Figure 3.1. Equilibrium between three interfacial tensions.

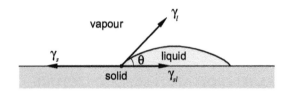

Figure 3.2. Interfacial tensions and wetting angle.

direction, real equilibrium is not maintained between the three phases. When $\gamma_s \geq \gamma_{sl} + \gamma_l$, θ is $0°$ and the solid is completely wetted by the liquid. Wettability and wetting angle are usually measured by the sessile drop method where the shape of a liquid drop on a substrate is defined.[4]

The wettability of a liquid on a substrate is an important property in a number of ceramic processes. With a decreased wetting angle, i.e. an increased wettability, densification is enhanced in liquid phase sintering and the bonding in brazing or soldering is improved.

3.1.2 Dihedral Angle

The angle at the junction between two grains (Figure 3.3) is referred to as the dihedral angle. If the interfacial tensions are in equilibrium at this junction,

$$\gamma_b = 2\gamma_{\alpha\beta} \cos\frac{\phi}{2} \tag{3.3}$$

is satisfied, where γ_b is the grain boundary energy of α, $\gamma_{\alpha\beta}$ the interfacial energy between α and β, and ϕ the dihedral angle. The dihedral angle is determined only by the interfacial energies and is independent of the pressures in the phases.[9] This means that the dihedral angle is constant, irrespective of the pressure of phase β. If β is a vapour phase, ϕ is larger than $120°$ because γ_s is usually higher than γ_b. In general, γ_s is 2–3 times higher than γ_b and ϕ is around $150°$. If the dihedral angle is constant and the junction edges are

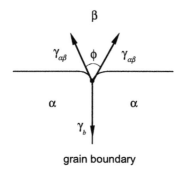

Figure 3.3. Interfacial tensions and dihedral angle.

randomly distributed in three dimensions, the measured angle with the maximum frequency is the true dihedral angle.[10,11]

3.2 SINGLE-PHASE MICROSTRUCTURES

In a single-phase material the microstructure is determined by the grain boundary energy which varies with the crystallographic orientations between adjacent grains. Figure 3.4 depicts schematically the energy of a symmetric tilt boundary with tilt angle θ. As shown in this figure, the grain boundary energy in polycrystals is not constant and local microstructures, which are determined by the tension equilibrium condition, can vary. However, for an equilibrium grain shape it can be assumed that the grain boundary energy is constant. This assumption is well accepted for a soap bubble structure. The equilibrium shape of soap bubbles (grains) is determined by two conditions:

(i) minimization of total interfacial area under interfacial tension equilibrium and
(ii) complete packing of space without voids.

Here, equilibrium shape means the shape of grains with the same shape and size.

   In two dimensions, the polygon that satisfies the above conditions is a hexagon. In three dimensions, the surface tension equilibrium condition requires that the corner of the grains is a point where six planes (grain boundaries) and four lines (grain edges) meet with each other. This geometry is realized by six soap films meeting at the centre point in a tetrahedral frame, as shown in Figure 3.5.[12] From Figure 3.5, the number of corners of a grain which satisfies this geometry is calculated to be $\sim$22.8. Therefore, there is no equilibrium polyhedron which satisfies the interfacial tension equilibrium condition. The polyhedra that have the corners close to 22.8 are a pentagonal dodecahedron and a tetrakaidecahedron. A pentagonal dodecahedron consists

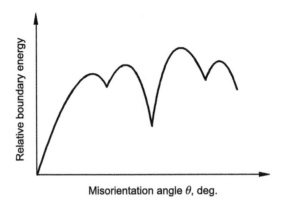

Figure 3.4. Schematic of the variation of grain boundary energy for symmetric tilt boundaries with tilt angle.

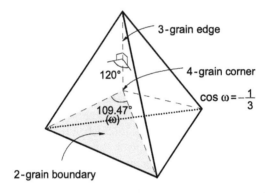

Figure 3.5. Soap film equilibrium in a tetrahedral frame.[12]

of twelve pentagons and has 20 corners, and a tetrakaidecahedron consists of six regular squares and eight regular hexagons, and has 24 corners, as shown in Figure 3.6. It can be said, therefore, that there is no equilibrium grain shape and no equilibrium microstructure in bulk polycrystals. However, when tetrakaidecahedra are packed in a body-centred cubic (bcc) lattice, the space is completely filled, although the interfacial tension equilibrium condition is not satisfied.[12–14] In a three-dimensional single-phase microstructure, four grains meet at a point (4-grain corner), three grains at an edge (3-grain edge) and two grains on a plane (2-grain boundary), as shown in Figure 3.5. The grains observed in real microstructure have many corners similar to that in Figure 3.5 and have curved grain boundaries.

However, all microstructures satisfy a topological relationship[12] represented by Euler's law which relates various dimensions in a geometrical figure. Let *C* (corner), *E* (edge), *P* (polygon) and *B* (body) be the numbers of

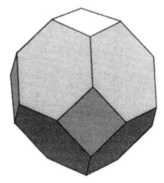

Figure 3.6. Shape of tetrakaidecahedron (a truncated octahedron).

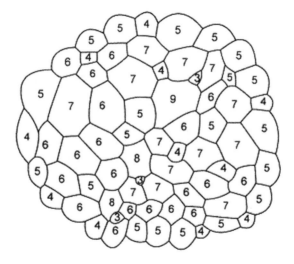

Figure 3.7. Illustration of topological relationship (Eq. (3.5)) in a network of polygons (a schematic microstructure).

0-, 1-, 2-, and 3-dimensional geometries, respectively. Euler's law states that

$$C - E + P - B = 1 \qquad (3.4)$$

holds. This equation is independent of the number and shape of any geometry. For a two-dimensional figure, $C - E + P = 1$ is satisfied.

We can describe the characteristics of a two-dimensional microstructure using Euler's law. In Figure 3.7, a typical two-dimensional microstructure, three polygons meet at a point. Therefore $3C = \Sigma nP_n + E_c$, where P_n is the number of polygons with n edges and E_c the number of partially shared corners in the periphery of the figure where only two polygons form a corner.

Unlike the corners, an edge forms when two polygons meet with each other. Therefore, $2E = \Sigma nP_n + E_b$, where E_b is the number of unshared edges in the periphery. When these equations are inserted in Eq. (3.4),

$$\sum (6 - n)P_n - E_b = 6 \tag{3.5}$$

Figure 3.7 exactly satisfies Eq. (3.5).

Let P be the total number of polygons and p_n the fraction of n-sided polygons. Equation (3.5) is then

$$\sum (6 - n)p_n = (E_b + 6)/P \tag{3.6}$$

For a microstructure with a large number of polygons, the right-hand side of Eq. (3.6) can be neglected and

$$\sum (6 - n)p_n = 0 \tag{3.7a}$$

$$4p_2 + 3p_3 + 2p_4 + 1p_5 \pm 0p_6 - 1p_7 - 2p_8 - \cdots - (n - 6)p_n = 0 \tag{3.7b}$$

This equation indicates that the average grain in a two-dimensional microstructure is a hexagon and that the polygons are distributed to satisfy the equation. For example, if a triangular grain is present in the microstructure, there must also be a grain with 9 sides or two grains with 7 and 8 sides. For grain boundaries with the same energy and hence a dihedral angle of 120°, a grain with more than 6 sides has, on average, concave boundaries and a grain with less than 6 sides, convex boundaries.

For a three-dimensional bulk microstructure, a relationship among polyhedra (B), polygons (P), edges (E) and points (C) can also be derived,[12] using Eq. (3.4), as

$$\sum (6 - n)P_n = 6(B + 1) \tag{3.8}$$

Let $\bar{n}$ be the average number of edges of all polygons, $\Sigma nP_n/P$, the following equations hold.

$$C/B = \bar{n}/[(6 - \bar{n}) - 6/P] \tag{3.9a}$$
$$P/B = 6/[(6 - \bar{n}) - \bar{n}/C] \tag{3.9b}$$
$$\bar{n} = 6[1 - (B + 1)/P] \tag{3.9c}$$
$$P - C = B + 1 \tag{3.9d}$$

If a bulk microstructure consists of tetrakaidecahedra, $P/B = 7$, $C/B = 6$ and $\bar{n} = 36/7$.[12]

3.3 **MULTIPHASE MICROSTRUCTURES**

Multiphase microstructure is determined, in general, by the local equilibrium among interfacial tensions. Figure 3.8 depicts schematically the local shape of three grains in contact. When local equilibrium in tension is maintained, the sine law (Eq. (3.1)) is satisfied. A second phase present at a junction of three such grains (called a triple junction) can have various shapes depending on the dihedral angle, as shown in Figure 3.9 for a two-dimensional microstructure.

In three dimensions, a second phase is elongated along three grain edges as the dihedral angle ϕ decreases and forms a continuous network when ϕ is equal to or less than $60°$.[13] Figure 3.10 shows schematically the distributions of a second phase for ϕ of less and more than $60°$. The second phase is isolated for $\phi > 60°$ but is continuous along three-grain edges for $\phi < 60°$. A real microstructure of a liquid phase sintered alloy with $\phi \approx 30°$ in Figure 3.11[15] is similar to the schematic microstructure in Figure 3.10(b). As the dihedral angle decreases, the grain boundary area decreases.

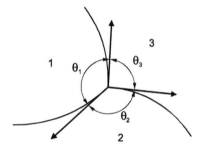

Figure 3.8. Schematic of interfacial tensions in equilibrium between three grains.

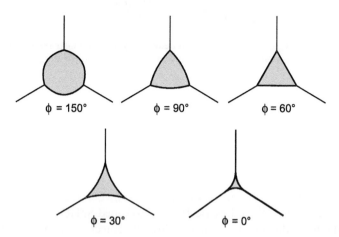

Figure 3.9. Distribution of a second phase with various dihedral angles in a two-dimensional microstructure.[12]

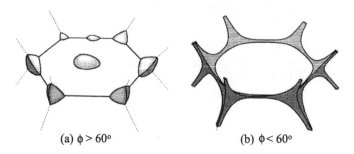

(a) $\phi > 60°$ (b) $\phi < 60°$

Figure 3.10. Schematic of three-dimensional distribution of a second phase with dihedral angle of (a) $>60°$ and (b) $<60°$.

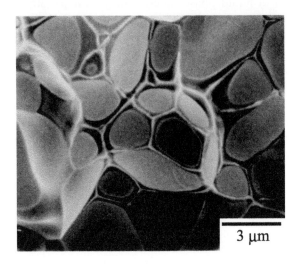

3 μm

Figure 3.11. Fracture surface of a W-Ni-Fe heavy alloy showing distribution of a Ni-Fe-W matrix phase between W(Ni,Fe) grains. 98W-1Ni-1Fe(wt%) sample sintered at 1460°C for 10 min in H_2.[15]

When the dihedral angle is $0°$, a liquid film is present at the grain boundary and all the grains are separated. In reality, however, some grain boundaries (low energy grain boundaries) can remain intact[16] because the grain boundary energy is not constant, as shown schematically in Figure 3.4. On the other hand, recent investigations show that the presence of a liquid film between grains can depend on the processing path of powder compacts although the dihedral angle is $0°$.[17,18] In the case of Nb_2O_5-Doped $SrTiO_3$, a preannealing of the sample in a reducing atmosphere below the eutectic temperature prevented the penetration of a liquid between grains when sintered above the eutectic temperature in the same atmosphere.[18] The dihedral angle of this sample was apparently $0°$, i.e. a liquid film was present between faceted grains when annealed above the eutectic temperature without the preannealing. Such a difference in liquid distribution between samples with different processing paths can be explained by a theoretical calculation which predicted the existence of two energy minima with and without a liquid film.[17]

The thickness of liquid films in ceramics was predicted and measured to be less than a few nm and to be nearly invariable with external pressure of an order of 10 atm.[17,19,20] The properties of the liquid film are considered to be different from those of the bulk liquid and to be close to those of the bulk solid.[21] According to a recent investigation,[22] however, a liquid film can form at dry grain boundaries and thicken during grain growth, as shown in Figure 3.12. With the growth of a $BaTiO_3$ single crystal into fine matrix $BaTiO_3$ grains ($\sim2\,\mu m$ in size) with $0.4\,mol\%$ excess TiO_2 in the matrix, a liquid film formed (Figure 3.12(b)) and thickened (Figure 3.12(c)) at originally dry grain boundaries (Figure 3.12(a)) between the single crystal and fine matrix grains at a temperature above the eutectic temperature. As the amount of excess TiO_2 increased, the film thickness increased considerably, $\sim12\,nm$ in the bi-layer sample with $1.0\,mol\%$ excess TiO_2 after annealing at $1350°C$ for 50 h. This kinetic formation and thickening of liquid films were attributed to an accumulation of impurities segregated at grain boundaries and penetration of the liquid present at triple junctions into grain boundaries during grain growth.

For a given dihedral angle and a given volume fraction of second phase (matrix), the grains assume shapes that have minimum interfacial energy.[23,24] Park and Yoon[24] calculated the interfacial energy of a grain as a function of matrix volume fraction and dihedral angle with respect to the grain boundary energy $(4\sqrt{2}l^2\gamma_b)$ of a rhombic dodecahedral grain with edge length l. The curves in Figure 3.13, which were calculated for a grain of constant volume, show that there is a specific fraction of the matrix volume needed to get minimum interfacial energy for a given dihedral angle of the system. When the dihedral angle is greater than $90°$, a fully dense polycrystal without a second phase has minimum interfacial energy since the interfacial energy increases monotonically as the volume fraction of the matrix increases. On the other hand, for a system with a dihedral angle less than $90°$, the minimum interfacial energy is achieved with a specific fraction of the matrix phase. Consider, for example, a single-phase polycrystal with bcc-packed rhombic dodecahedral grains in contact with a liquid that forms a dihedral angle of $30°$. In this case, liquid penetrates the polycrystal until its volume fraction is $\sim18\%$ and the grains become rounded although their connectivity is maintained through the grain boundaries. When the liquid volume fraction is less than 18%, the grains are less rounded than those in the compact with 18% liquid. On the other hand, for more than 18% liquid, the excess liquid over 18% remains outside the compact. The curves calculated beyond the minima with respect to matrix volume fraction for $\phi > 0°$ show the interfacial energies obtained when the excess matrix is pushed into the compact until the grains become spherical.

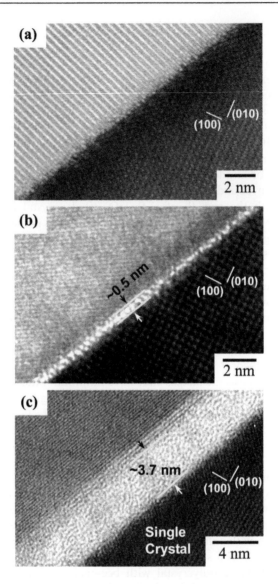

Figure 3.12. HREM images of the boundaries between a single crystal and a fine matrix grain. (100) single crystal/polycrystal bi-layer samples with 0.4-mol%-TiO_2 addition annealed at 1350°C for (a) 5, (b) 20, and (c) 50 h in air after H_2-treatment at 1250°C for 10 h.[22]

The discussion so far indicates that, when the matrix fraction is not equal to that associated with the minimum interfacial energy for a given dihedral angle, there is a driving force for grains to attain a shape with minimum interfacial energy. In other words, there is a certain pressure, i.e. an effective pressure towards a shape with minimum interfacial energy, present in and at the

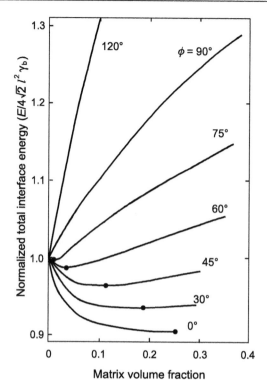

Figure 3.13. Calculated variation of total interfacial energy, E, with the matrix volume fraction (at constant grain volume) for various dihedral angles.[24] The minimum E values are shown by filled circles.

compact. In the case of a compact in solid state sintering where a second phase is vapour, the effective pressure is the sintering pressure. The effective pressure can be calculated following a similar procedure to that for the calculation of capillary pressure in Section 2.3.1 where the work done by an infinitesimal change in volume under an effective pressure is equal to the change in total interfacial energy of the system. For a constant volume of solid and a variable volume of matrix, the effective pressure P_e is expressed as[24]

$$P_e = -\frac{(1-f_m)^2}{V_g}\left[\frac{\partial E}{\partial f_m}\right]_{V_g} \tag{3.10}$$

where f_m is the volume fraction of the matrix, V_g the volume of a grain, and E the total interfacial energy of a grain. Equation (3.10) indicates that the effective pressure is inversely proportional to the grain size and is proportional to the slope of the calculated curve in Figure 3.13.

PROBLEMS

1.1. Derive the energy change when a cube-shaped powder of edge l is densified without grain growth. From the calculated energy change, discuss conditions to improve sinterability. The specific surface energy of the powder is γ_s and the specific grain boundary energy γ_b.

1.2. You have a steel powder and a steel sheet with the same chemical composition. Draw schematically and explain the changes in Gibbs free energy with processing time for the following processes: (a) sintering of the steel powder, (b) cold rolling of the steel sheet, and (c) heat treatment (annealing) of the steel sheet. Assume that the sintering and heat treatment temperatures are the same, and their periods of time are also the same. The heat treatment is assumed to follow a quenching and the sintering to follow a furnace cooling.

1.3. Derive Eq. (2.15) $[\sigma_{ij} = \delta_{ij}\gamma + (d\gamma/d\varepsilon_{ij})]$.

1.4. Two elastic balloons that contain different amounts of air are separated by a valve in a glass tube, as shown in Figure P1.4. When the valve is opened, what will happen?

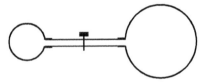

Figure P1.4

1.5. What is the pressure for shrinkage of a spherical pore entrapped within a single crystal sphere? What are your assumptions for the answer?

1.6. Pores with a radius of $5\,\mu m$ containing insoluble gases of $1\,atm$ pressure are entrapped within a glass with a relative density of 0.90. What are the equilibrium size of the pores and the final density of the glass? γ_s is assumed to be $0.5\,J/m^2$.

1.7. Consider an isolated spherical pore with a radius of $3\,\mu m$ which was entrapped within a Cu powder compact during pressing at $20°C$. Answer whether the pore expands or shrinks at $1000°C$. Assume the initial pressure of insoluble gases entrapped within the pore to be $10^5\,N/m^2$ and the surface energy of Cu at $1000°C$ to be $1.4\,J/m^2$.

1.8. Prove that the energy increase caused by the formation of a particle with a radius of r from the bulk state is $4\pi r^2 \gamma_s$.

1.9. Calculate the size of a Cu sphere where the surface energy is equal to the elastic strain energy due to its curvature. Assume that the surface energy γ_s of Cu is 1.4 J/m^2 and its compressibility $\kappa = 7.1 \times 10^{-12}$ m^2/N. Discuss the result.

1.10. When ten water drops of 1 μm radius stick together to make a large drop, what is the energy change? The compressibility of water is 4.5×10^{-10} m^2/N and the surface energy 7.3×10^{-2} J/m^2.

1.11. Draw a figure showing the change in equilibrium vapour pressure with the radius of curvature of a material (from 0.01 μm to 10 μm) at 1000°C. Assume the molar volume and the surface energy of the material to be 1×10^{-5} m^3 and 1 J/m^2, respectively.

1.12. When a tube is inserted into a liquid, the liquid level in the tube varies with the wetting angle θ, the tube radius r, liquid density ρ and surface tension of the liquid γ_l. Derive the equation of the liquid level height using two different concepts: capillary pressure of the liquid in the tube and surface tension of the liquid.

1.13. When the molar volumes of α and β phases are the same, Eq. (2.28) suggests that the pressure P^α in an α particle with a radius of r embedded in β becomes infinite. Is this consequence reasonable? Discuss.

1.14. Prove that the dihedral angle between two grains is independent of external isostatic pressure acting on the grain surfaces.

1.15. What is the equilibrium shape of an entrapped pore within a crystal whose equilibrium shape is a cube in air?

1.16. The microstructure in Figure P1.16 consists of α (dark area) and β (light area) phases. Explain a possible method for measuring the ratio of the grain boundary energies between α/α and β/β phases, $\gamma_{\alpha/\alpha}/\gamma_{\beta/\beta}$, and describe the assumptions made for the measurement. Which grain boundary has a higher energy?

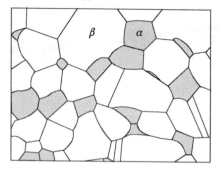

Figure P1.16

1.17. Derive Eqs (3.8) and (3.9).

1.18. When a polycrystal is immersed in a liquid which is in chemical equilibrium with the solid, the liquid can penetrate along the grain boundaries according to the dihedral angle and form a solid–liquid two-phase micro-structure. However, if the boundary energy is not constant but varies with grain boundary orientation, some boundaries with low energies may not be penetrated by the liquid even though the apparent dihedral angle is 0°. Assuming the nearest neighbour grains on a two-dimensional plane to be six, a researcher estimated the fraction of the unpenetrated grain boundary area in three dimemsions to be (*number of grain boundaries/ number of grains* × 6) × 100(%). Discuss the validity of his estimation.

1.19. Let γ_b be the grain boundary energy of α and γ_{sl} be the interfacial energy between an α grain and the matrix. Draw schematically and explain the variation of γ_{sl}/γ_b with dihedral angle ϕ.

1.20. Calculate the differences in total interfacial energy when a spherical second-phase particle with a radius of r is located at a 2-grain boundary, a 3-grain edge and a 4-grain corner. Assume that the dihedral angle is 180° although the grain boundary energy γ_b is finite.

1.21. Describe possible microstructural changes with annealing time (from 0 h to ∞ h) of an oxide polycrystal which was fully densified by hot pressing in a vacuum. Assume that the dihedral angle is 75°.

1.22. To prepare a bulk $YBa_2Cu_3O_{7-x}$(123) superconductor with high critical current density, a melt-texturing technique, which consists of a peritectic reaction between Y_2BaCuO_5(211) and an oxide melt, is often utilized. The 123 grains, which are formed through the peritectic reaction between the 211 phase and the oxide melt, grow preferentially in a <100> direction and often trap isolated pores in the melt. In a system containing $BaCeO_3$, the shape of the entrapped pores within the 123 grains is a peculiar one, as shown in Figure P1.22.[25] Explain possible causes of the formation of the crystallographically aligned pores with an elongated shape.

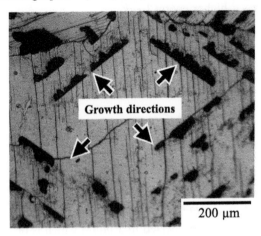

Figure P1.22

1.23. A liquid phase sintered body has, in general, a microstructure with well-distributed grains in a matrix. For a liquid phase sintered body containing gas bubbles, draw schematically the shape of a gas bubble on a grain for a wetting angle θ larger and smaller than 90°, respectively. For both cases, do you expect a difference in mechanical properties, for example, tensile strength? Explain.

1.24. In general, grain boundary energy varies with crystallographic orientation between grains. Provided that a liquid penetrated into all of the grain boundaries of a polycrystal, draw schematically and explain grain boundary energy versus grain boundary orientation and solid/liquid interfacial energy versus grain boundary orientation.

1.25. Consider a polycrystal with an isotropic grain structure. If the polycrystal is disintegrated into isotropic rounded grains in liquid A and anisotropic angular grains in liquid B, what would be the variation of γ_b and γ_{sl} with the tilt angle of grain boundaries? Explain.

1.26. Consider two polycrystals with and without a small amount of liquid, say 1 vol%, with a dihedral angle of 0°. Describe the difference in shape of a boundary between a large and a small grain for both of the polycrystals, if any. The grain boundary energy and solid/liquid interfacial energy are assumed to be invariable with grain boundary orientation.

1.27. Consider a system where α–β–liquid are in equilibrium at a certain temperature, and the dihedral angles of α and β are 30° and 5°, respectively. When a dense α–liquid compact and a dense β–liquid compact with 5 vol% of liquid each are in contact at the temperature, what will be the change in microstructure directly after contact, after a short period of time and after a long period of time of contact? Assume that the wetting angle is 0°, the molar volumes of α and β are the same and the diffusivities of different atoms in the liquid are the same.

1.28. Park and Yoon calculated the total interfacial energy of two-phase systems, as shown in Figure 3.13.[24] According to their calculation, for a system with a dihedral angle ϕ, $0° < \phi < 90°$, the interfacial energy is a minimum at a finite volume fraction of the matrix phase. If the matrix volume fraction in a powder compact is larger than that for the minimum interfacial energy, what will happen during sintering of the compact? Under what condition is the total interfacial energy calculated for the liquid volume fraction larger than that for the minimum interfacial energy? What does the right end of the calculated curves mean?

GENERAL REFERENCES FOR SINTERING SCIENCE

S1. Jones, W. D., *Fundamental Principles of Powder Metallurgy*, Edward Arnold, London, 1960.

S2. Thümmler, F. and Thomma, W., Sintering processes, *Metall. Review*, Vol. 12, 69–108, 1967.

S3. Geguzin, J. E., *Physik des Sinterns*, VEB Deutscher Verlag für Grundstoffindustrie, Leipzig, 1973.

S4. Lenel, F. V., *Powder Metallurgy: Principles and Applications*, MPIF, Princeton, 1980.

S5. Exner, H. E. and Arzt, E., Sintering processes, in *Physical Metallurgy* (3rd edition), Chap. 30, R. W. Cahn and P. Haasen (eds), Elsevier Science Publishing, Amsterdam, 1885–912, 1983.

S6. German, R. M., *Liquid Phase Sintering*, Plenum Press, New York, 1985.

S7. Rahaman, M. N., *Ceramic Processing and Sintering* (1st and 2nd editions), Marcel Dekker, New York, 1995 and 2003.

S8. German, R. M., *Sintering Theory and Practice*, John Wiley & Sons, New York, 1996.

S9. Kingery, W. D., Bowen, H. K. and Uhlmann, D. R., *Introduction to Ceramics* (2nd edition), John Wiley & Sons, New York, 1976.

S10. Chiang, Y.-M., Birnie III, D. and Kingery, W. D., *Physical Ceramics*, John Wiley & Sons, New York, 1997.

S11. Trivedi, R. K., Theory of capillarity, in *Lectures on the Theory of Phase Transformations*, Chap. 2, H. I. Aaronson (ed.), AIME, New York, 51–81, 1975.

S12. Murr, L. E., *Interfacial Phenomena in Metals and Alloys*, Addison-Wesley, London, 1975.

S13. Martin, J. W. and Doherty, R. D., *Stability of Microstructure in Metallic Systems*, Cambridge University Press, Cambridge, 1976.

S14. Howe, J. M., *Interfaces in Materials*, John Wiley & Sons, New York, 1997.

S15. Lupis, C. H. P., *Chemical Thermodynamics of Materials*, Elsevier Science Publishing, New York, 1983.

REFERENCES

1. Sheppard, L. M., Maintaining competitiveness in the age of materials, *Am. Ceram. Soc. Bull.*, **68**, 2038–39, 1989.
2. General textbooks on thermodynamics and interfaces, such as General References from S11 to S15.
3. Mullins, W. W., Solid surface morphologies governed by capillarity, in *Metal Surfaces: Structure, Energetics and Kinetics*, ASM, Metals Park, Ohio, 17–66, 1963.
4. Murr, L. E., *Interfacial Phenomena in Metals and Alloys*, Addison-Wesley, London, 87–164, 1975.
5. Kingery, W. D., Bowen, H. K. and Uhlmann, D. R., *Introduction to Ceramics* (2nd edition), John Wiley & Sons, New York, 177–216, 1976.
6. Chiang, Y.-M., Birnie III, D. and Kingery, W. D., *Physical Ceramics*, John Wiley & Sons, New York, 351–71, 1997.
7. Cahn, J. W. and Hoffman, D. W., A vector thermodynamics for anisotropic interfaces II. Curved and faceted surfaces, *Acta Metall.*, **22**, 1205–14, 1974.
8. King, A. H., Equilibrium at triple junctions under the influence of anisotropic grain boundary energy, in *Grain Growth in Polycrystalline Materials*, H. Weiland, B. L. Adams and A. D. Rollett (eds) TMS, Warrendale, PA, 333–38, 1988.
9. Rice, J. R. and Chuang, T.-J., Energy variations in diffusive cavity growth, *J. Am. Ceram. Soc.*, **64**, 46–53, 1981.
10. Harker, D. and Parker, E. R., Grain shape and grain growth, *Trans. ASM*, **34**, 156–201, 1945.
11. Ball, C. J., Estimation of the dihedral angle between spherical grains from measurements in a plane section, *Trans. Brit. Ceram. Soc.*, **65**, 41–49, 1966.
12. Smith, C. S., Some elementary principles of polycrystalline microstructure, *Metall. Reviews*, **9**, 1–48, 1964.
13. Smith, C. S., Grains, phases and interfaces: an interpretation of microstructure, *Trans., AIME*, **175**, 15–51, 1948.
14. Chadwick, G. A., *Metallography of Phase Transformations*, Butterworth & Co., London, 160–82, 1972.
15. Park, H.-H., Kang, S.-J. L. and Yoon, D. N., An analysis of the surface menisci in a mixture of liquid and deformable grains, *Metall. Trans. A.*, **17A**, 325–30, 1986.
16. Kim, S. S. and Yoon, D. N., Coarsening of mo grains in the molten Ni-Fe matrix of a low volume fraction, *Acta Metall.*, **33**, 281–86, 1985.
17. Ackler, H. D. and Chiang, Y.-M., Effect of initial microstructure on final intergranular phase distribution in liquid phase sintered ceramics, *J. Am. Ceram. Soc.*, **82**, 183–89, 1999.

18. Chung, S.-Y. and Kang, S.-J. L., Intergranular amorphous films and dislocation-promoted grain growth in $SrTiO_3$, *Acta Mater.*, **51**, 2345–54, 2003.
19. Clarke, D. R., On the equilibrium thickness of intergranular glass phases in ceramic materials, *J. Am. Ceram. Soc.*, **70**, 15–22, 1987.
20. Clarke, D. R., Shaw, T. M., Philipse, A. P. and Horn, R. G., Possible electrical double-layer contribution to the equilibrium thickness of intergranular glass films in polycrystalline ceramics, *J. Am. Ceram. Soc.*, **76**, 1201–204, 1993.
21. Tu, K.-N., Mayer, J. W. and Feldman, L. C., *Electronic Thin Film Science for Electrical Engineers and Materials Scientists*, Macmillan Publishing Co., New York, 246–80, 1992.
22. Choi, S.-Y., Yoon, D. Y. and Kang, S.-J. L., Kinetic formation and thickening of intergranular amorphous films at grain boundaries in $BaTiO_3$, *Acta Mater.*, **52**, 3721–26, 2004.
23. Beere, W., A unifying theory of the stability of penetrating liquid phases and sintering pores, *Acta Metall.*, **23**, 131–38, 1975.
24. Park, H. H. and Yoon, D. N., Effect of dihedral angle on the morphology of grains in a matrix phase, *Metall. Trans. A.*, **16A**, 923–28, 1985.
25. Kim, C.-J., Hong, G.-W. and Kang, S.-J. L., Entrapment of elongated and crystallographically aligned pores in $YBa_2Cu_3O_{7-x}$ melt-textured with $BaCeO_3$ addition, *J. Mater. Res.*, **14**, 1707–10, 1999.

PART II
SOLID STATE SINTERING MODELS
AND DENSIFICATION

Solid state sintering is usually divided into three overlapping stages — initial, intermediate and final. Figure 4.1 schematically depicts the typical densification curve of a compact through these stages over sintering time. The initial stage is characterized by the formation of necks between particles and its contribution to compact shrinkage is limited to 2–3% at most. During the intermediate stage, considerable densification, up to ~93% of the relative density, occurs before isolation of the pores. The final stage involves densification from the isolated pore state to the final densification. For each of these three stages, simplified models are typically used: the two-particle model for the initial stage, the channel pore model for the intermediate stage, and the isolated pore model for the final stage. Although all models ignore grain growth during sintering, they do provide a means of analysing the densification process and evaluating the effects of various processing parameters. Chapter 4 deals with sintering mechanisms and kinetics of initial stage sintering. In Chapter 5, densification models and theories at intermediate and final stage sintering are described. Topics of pressure-assisted sintering and constrained sintering are also included in Chapter 5.

PART II
SOLID STATE SINTERING MODELS
AND DENSIFICATION

4

INITIAL STAGE SINTERING

4.1 TWO-PARTICLE MODEL

4.1.1 Geometrical Relationships

The sintering of powder compacts with complex-shaped particles of different sizes cannot be explained in a simple manner. However, if spherical particles of the same size are assumed, the sintering of powder compacts can be represented as the sintering between two particles, as shown in Figure 4.2. In early studies, however, a sphere/plate geometry was also used for the explanation of initial stage sintering.[1] Since the driving force behind sintering is largely determined by neck geometry and size, an examination of the geometrical relationship around the neck is first needed.

Figure 4.2 shows two geometrical models for two spherical particles: one without shrinkage (a) and the other with shrinkage (b). In Figure 4.2(a), the distance between the particles does not change but the neck size increases as the sintering time increases. In the model with shrinkage (Figure 4.2(b)), the neck size increases with an increased sintering time by material transport between the particles and hence shrinkage results. If the dihedral angle between the particles is 180° and the grain size does not change during sintering, the radius of neck curvature r, neck area A and neck volume V are respectively

$$r \approx \frac{x^2}{2a} \tag{4.1}$$

$$A \approx 2\pi x \, 2r = \frac{2\pi x^3}{a} \tag{4.2}$$

and

$$V = \int A \mathrm{d}x = \frac{\pi x^4}{2a} \tag{4.3}$$

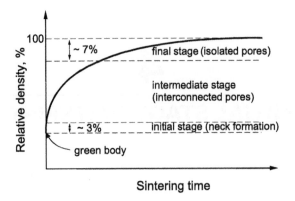

Figure 4.1. Schematic showing the densification curve of a powder compact and the three sintering stages.

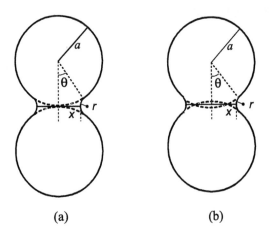

(a)   (b)

Figure 4.2. Two-particle model for initial stage sintering (a) without shrinkage and (b) with shrinkage.

for the geometry without shrinkage (Figure 4.2(a)). Here, a is the particle radius and x the neck radius.

For the geometry with shrinkage (Figure 4.2(b)),

$$r \approx \frac{x^2}{4a} \tag{4.4}$$

$$A \approx \frac{\pi x^3}{a} \tag{4.5}$$

and

$$V \approx (\pi x^2 \, 2r)/2 = \frac{\pi x^4}{4a} \tag{4.6}$$

Comparing Eqs (4.1)–(4.3) with Eqs (4.4)–(4.6), it is seen that if shrinkage occurs the values are respectively one-half of those without shrinkage.

If the dihedral angle is less than $180°$ as in real systems, the values of r are larger than those in Eq. (4.1) and Eq. (4.4). In the absence of shrinkage, r is calculated to be[2]

$$r = \frac{x^2}{2a\left[1 - \left(\frac{x}{a}\right)\sin\frac{\phi}{2} - \cos\frac{\phi}{2}\right]}$$

$$\approx \frac{x^2}{2a\left(1 - \cos\frac{\phi}{2}\right)} \qquad \left(\frac{x}{a} \ll 1\right) \qquad (4.7)$$

4.1.2 Driving Force and Mechanisms of Sintering in the Two-Particle Model

The driving force of sintering appears as differences in bulk pressure, vacancy concentration and vapour pressure — parallel phenomena — due to differences in surface curvature of the particles. For the geometry in Figure 4.2, the pressure difference ΔP is

$$\Delta P = P_a - P_r = \gamma_s\left(\frac{2}{a} + \frac{1}{r} - \frac{1}{x}\right)$$

$$\cong \frac{\gamma_s}{r} \qquad (a \gg x \gg r) \qquad (4.8)$$

the vacancy concentration difference ΔC_v is

$$\Delta C_v = C_{v,\infty}\frac{V'_m}{RT}\frac{\gamma_s}{r} \qquad (4.9)$$

and the vapour pressure difference Δp is

$$\Delta p = p_\infty\frac{V_m}{RT}\frac{\gamma_s}{r} \qquad (4.10)$$

Here, γ_s is the specific surface energy of the solid (solid surface energy), V'_m the molar volume of vacancies and V_m the molar volume of the solid. In general, V'_m is not the same as V_m because of the relaxation of atoms around the vacancy.[3]

The differences in bulk pressure (Eq. (4.8)), vacancy concentration (Eq. (4.9)) and vapour pressure (Eq. (4.10)) induce material transport. Table 4.1 lists the major mechanisms of material transport and their related parameters. Figure 4.3 illustrates the paths of material transport for the transport mechanisms listed in Table 4.1. The material transport due to the difference in interface curvature occurs under the parallel actions of various mechanisms. The dominant mechanism, however, can vary depending on, for example,

Table 4.I. Material transport mechanisms during sintering

Material transport mechanism	Material source	Material sink	Related parameter
1. Lattice diffusion	Grain boundary	Neck	Lattice diffusivity, D_l
2. Grain boundary diffusion	Grain boundary	Neck	Grain boundary diffusivity, D_b
3. Viscous flow	Bulk grain	Neck	Viscosity, η
4. Surface diffusion	Grain surface	Neck	Surface diffusivity, D_s
5. Lattice diffusion	Grain surface	Neck	Lattice diffusivity, D_l
6. Gas phase transport			
6.1. Evaporation/ condensation	Grain surface	Neck	Vapour pressure difference, Δp
6.2. Gas diffusion	Grain surface	Neck	Gas diffusivity, D_g

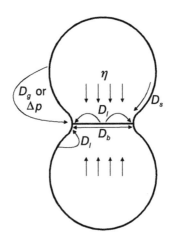

Figure 4.3. Material transport paths during sintering.

particle size, neck radius, temperature and time for a given system (see Section 4.3).

Some of these material transport mechanisms contribute to densification and shrinkage while others do not. The interparticle distance can be reduced only by bulk material flow via viscous flow or by material transport from the grain boundary via atom movement. When material comes to the neck from the particle surface, interparticle distance is not reduced but the neck size is increased by redistribution of material. Therefore, the grain boundary is the source of material transport for densification and shrinkage in crystalline powder compacts.

The importance of the grain boundary as a source of material for sintering was not recognized when various sintering models were proposed in the

1940s;[1,4,5] this significance was recognized in the 1950s by a couple of model experiments.[6,7] Alexander and Balluffi[7] used a Cu wire wound Cu spool to show that only pores attached to grain boundaries shrank while those entrapped within grains did not. Further experimental verification of various sintering mechanisms was also made in the 1960s.[8,9,S2,S3]

In discussions of sintering mechanisms, there has been considerable debate as to the possibility of plastic deformation of the crystalline particles, similar to a bulk flow of amorphous materials, during sintering.[8–10] Although this can be a major mechanism in pressure-assisted sintering (see Section 5.6), it is unimportant under atmospheric sintering, as Kuczynski et al. showed.[8] Furthermore, calculation shows that the capillary force in the neck region is usually far less than the stress required to generate dislocations, $2\mu b/l$,[11] where μ is the shear modulus, b the Burgers vector and l the distance between the pinned tips for dislocation generation. Plastic deformation may occur only when the particles are in point contact at the very beginning of sintering.[12] (See Section 4.3.) On the other hand, based on a measurement of dislocation density in the neck region between particles, Schatt et al. proposed that the dislocation density varied with sintering time, and this variation in dislocation density critically affected the sintering.[13] However, there has been little experimental confirmation to support dislocation-activated sintering.

4.1.3 Atom Diffusion and Diffusion Equations

Diffusion is the most important sintering mechanism. The diffusion mechanism is related to the movement of atoms under a difference in vacancy concentration. Atom movement itself may be interpreted physically in two ways: namely, atom movement as a result of vacancy diffusion under a difference (gradient) in vacancy concentration and movement of the atoms themselves under a difference (gradient) in stress.

In terms of vacancy movement, the vacancy flux, $J_{vac.}$ is expressed as

$$J_{vac.} = -D_v \nabla C_v$$
$$= -\frac{D_v C_{v,\infty} V'_m}{RT}(\Delta P)\frac{1}{L} \qquad (4.11)$$

where D_v is the vacancy diffusion coefficient, C_v the vacancy concentration per unit volume, $C_{v,\infty}$ the equilibrium vacancy concentration in the material with a flat surface, and L the diffusion distance. In terms of atom movement, the atom flux J_{atom} is expressed as[14]

$$J_{atom} = -C_a B_a \nabla(\mu_a - \mu_v) \qquad (4.12)$$

where C_a is the atom concentration per unit volume, B_a the atom mobility, μ_a the chemical potential of the atom, and μ_v the chemical potential of the vacancy. Equation (4.12) is applicable under the assumption that B_a is invariable with the location of the atom. Berrin and Johnson[15] suggested that the vacancy does not move on average if the formation and annihilation of vacancies occurs freely, and the vacancy concentration in any region is in local equilibrium. Under these conditions, there would be no difference in the chemical potential of a vacancy in different locations and effectively no movement of vacancies. Equation (4.12) can then be written as

$$J_{atom} = -C_a B_a \nabla \mu_a$$

$$= -C_a \frac{D_a}{RT} \nabla \sigma V_m$$

$$= -\frac{D_a}{RT} \nabla \sigma = -\frac{D_a}{RT}(\Delta P)\frac{1}{L} \tag{4.13}$$

where D_a is the atom diffusion coefficient and σ the pressure. This equation is the same as Eq. (4.11) because $D_v C_v = D_a C_a$.

Therefore, we can use either Eq. (4.11) or Eq. (4.13) to describe the kinetics of sintering that occurs by diffusion. However, the concept of atom movement by chemical potential difference of atoms must be more general because the vacancy movement mechanism can erroneously describe the sintering kinetics of dopant-added compounds.[16] According to Johnson,[16] adding a dopant to create vacancies of slowly moving species results in a considerable reduction in the concentration gradient of the vacancy if the dopant is uniformly distributed, irrespective of the presence of pores (this assumption, however, may be questionable). Therefore, the vacancy movement mechanism may not be appropriate to explain the effect of dopant addition. On the other hand, in the atom movement mechanism, the effect of the dopant addition appears as an increase in atom mobility; this explains well the sintering kinetics of dopant-added materials. Sintering by diffusion should be interpreted as material transport in a chemical potential gradient rather than that in a vacancy concentration gradient.

4.1.4 General Features of the Sintering Model and its Limitations

A basic assumption in sintering is that the particles are sintered under a quasi-equilibrium state. This means that in the case of diffusion, the diffusion gradient is in a steady state and the time to achieve the steady state is negligible compared with that of a change in particle geometry. Therefore, atoms in any location are locally in equilibrium with a given capillary stress. Under this assumption, sintering kinetics is not controlled by the equilibration reaction and its kinetics at the interface but by atom movement. (This assumption does not apply to the evaporation/condensation mechanism described in Section 4.2.6.)

Another assumption is that the stress at the grain boundary is distributed so that atoms at the grain boundary come out to the neck uniformly. Physically, this assumption is based on the fact that no pores form at the grain boundary even though atoms at the boundary are transported to the neck. Exner and Bross calculated a parabolic distribution of stress, compressive in the centre region and tensile in the surface region of the neck, for a two-wire model.[17]

In sintering, the grain boundary and the surface of the particles are the source and sink of atoms. Dislocations can also be the source and sink of atoms; however, this contribution is low and not usually considered. The grain boundary is usually assumed to be a perfect source and sink where no energy is required for atoms to attach or detach. However, this assumption is hard to accept in real systems where the grain boundary itself can have a crystallographically quite well arranged structure.[18–20] Under this situation, an excess energy would be required for atom detachment or attachment.[21]

The grain boundary as an imperfect source and sink of atoms may be clearly appreciated when the boundary is faceted. A recent investigation on the sintering of $BaTiO_3$[22] showed that the densification and grain growth were greatly reduced when a structural transition of the grain boundary from rough to faceted occurred. This result indicates that an excess energy is needed to remove atoms from atomically flat boundaries. Another cause of the excess energy for atom removal from grain boundaries can be solute atoms or second phase particles at the grain boundaries.[21] If a dislocation at the grain boundary affects atom or vacancy movement, these atoms or particles can impede dislocation movement and reduce the driving force for atom movement (see Section 10.1). Therefore, in reality, the assumption of a perfect atom source and sink is not satisfied. Nevertheless, the assumption has usually been accepted in the development of sintering kinetics and theories.

4.2 SINTERING KINETICS

As explained above, sintering kinetics in the two-particle model may vary considerably according to the geometry and stress distribution assumed.[1,2,5] Although many different calculations of neck growth kinetics have been published, the precision of the different equations is debatable. Thus, the numerical constants in neck growth equations do not have absolute meaning while the dependence of neck size on sintering time has meaning within a physically acceptable limit. To discuss the neck growth during the initial stage sintering, neck size is assumed to be much smaller than particle size ($x/a < {\sim}0.2$ in Figure 4.2) and the angle θ between the neck centre and neck surface is assumed to be much less than 1 ($\theta \ll 1$). When material comes from the particle surface to the neck (Figure 4.2(a)), the particle surface near the neck must recede from the original rounded surface. When this phenomenon, called under-cutting,[23] occurs, the driving force for sintering is reduced.

4.2.1 Lattice Diffusion from Grain Boundaries

The lattice diffusion of atoms from the grain boundary to the neck allows the boundary to act as a site for vacancy annihilation. This concept of the role of the grain boundary is similar to that occurring during Nabarro–Herring creep.[24,25] The idea of atom movement from a grain boundary under compressive stresses to another grain boundary under tensile stresses during creep is the same as that of vacancy movement in the reverse direction. If lattice diffusion of atoms from the grain boundary to the neck surface is to occur, the neck region must be under tensile stresses and the grain boundary under compressive stresses. In this regard it would be reasonable to consider that a stress gradient is present along the grain boundary from the neck centre to the neck surface, as Exner and Bross calculated.[17] By this sintering mechanism, not only neck growth but also a centre-to-centre approach of the two particles, i.e. shrinkage, occurs because material is removed from the contact area of the particles. The neck growth and shrinkage kinetics can be derived in the following way:

Neck growth

From $dV/dt = JAV_m$ and the geometry in Figure 4.2(b),

$$\frac{\pi x^3}{a}\frac{dx}{dt} = \frac{D_l}{RT}\nabla\sigma\, A\, V_m$$

$$\approx \frac{D_l}{RT}\left(\frac{\gamma_s}{r}\frac{1}{x}\right)\frac{\pi x^3}{a} V_m$$

$$\therefore x^4 = \frac{16 D_l\gamma_s V_m a}{RT}t \tag{4.14}$$

Shrinkage

$$\frac{\Delta l}{l} = \frac{r}{a} = \frac{x^2}{4a^2} = \left(\frac{D_l\gamma_s V_m}{RTa^3}\right)^{1/2} t^{1/2} \tag{4.15}$$

where D_l is the lattice diffusion coefficient and l the sample size. Through the lattice diffusion mechanism, material can come to the neck surface not only from the grain boundary but also from the particle surface. The latter, however, does not contribute to the shrinkage (see Section 4.2.5).

4.2.2 Grain Boundary Diffusion from Grain Boundaries

In some respects material transport from the grain boundary to the neck by grain boundary diffusion is similar to diffusional creep by grain boundary diffusion (so-called Coble creep[26]).

Neck growth

$$\frac{dV}{dt} = \frac{\pi x^3}{a}\frac{dx}{dt} = \frac{D_b}{RT}\frac{\gamma_s}{r}\frac{1}{x}2\pi x\delta_b V_m$$

$$\therefore x^6 = \frac{48 D_b \delta_b \gamma_s V_m a^2}{RT}t \tag{4.16}$$

Shrinkage

$$\frac{\Delta l}{l} = \frac{r}{a} = \left(\frac{3 D_b \delta_b \gamma_s V_m}{4 R T a^4}\right)^{1/3} t^{1/3} \tag{4.17}$$

where D_b is the grain boundary diffusion coefficient and δ_b the diffusion thickness of grain boundary diffusion. In this case the material transported to the neck surface by grain boundary diffusion should be redistributed via another mechanism. If the redistribution of material is not fast enough compared to the material transport by grain boundary diffusion, this secondary redistribution may control the neck growth. (See Problem 2.13.)

4.2.3 Viscous Flow

The viscous flow mechanism, which was first proposed by Frenkel,[4] can be operative in the sintering of viscous materials like glass. If the material follows the behaviour of a Newtonian fluid, the neck growth and shrinkage kinetics are expressed as follows.

Neck growth

$$\dot{\varepsilon} = \frac{1}{h}\frac{dh}{dt} = \frac{1}{\eta}\Delta\sigma = \frac{1}{\eta}\frac{\gamma_s}{r}$$

$$\therefore dh = \frac{1}{\eta}\frac{h}{r}\gamma_s dt \approx \frac{1}{\eta}\gamma_s dt$$

$$\therefore h \approx \frac{x^2}{4a} = \frac{\gamma_s}{\eta}t$$

$$\therefore x^2 = \frac{4\gamma_s a}{\eta}t \tag{4.18}$$

Shrinkage

$$\frac{\Delta l}{l_o} = \frac{h}{a} \approx \frac{\gamma_s}{\eta a}t \tag{4.19}$$

where η is the viscosity of the material and h the penetration depth of one particle into another which is approximately r in Figure 4.2(b).

4.2.4 Surface Diffusion from Particle Surfaces

Sintering by surface diffusion occurs via atom movement on the surface of the spheres from the sphere surface to the neck surface. In this case a stress gradient derived from the capillary pressure on the neck surface may be assumed to be present up to the region within a distance equal to the neck curvature radius. (This assumption is different to that of the stress gradient for lattice or grain boundary diffusion from the grain boundary.) This means that there is no stress gradient beyond a distance equal to the neck curvature radius and that neck growth is controlled by atom transport by surface diffusion from this region. The material transport by this mechanism results in no shrinkage.

Neck growth

$$\frac{dV}{dt} = \frac{2\pi x^3}{a}\frac{dx}{dt} = JAV_m = \frac{D_s}{RT}\frac{\gamma_s}{r}\frac{1}{r}(2\pi x\, 2\delta_s)V_m$$

$$\therefore x^7 = \frac{56 D_s \delta_s \gamma_s V_m a^3}{RT} t \tag{4.20}$$

where D_s is the surface diffusion coefficient and δ_s is the diffusion thickness of the surface diffusion.

4.2.5 Lattice Diffusion from Particle Surfaces

As the source of material is the particle surface, there is no shrinkage even though the neck growth occurs by lattice diffusion. Under the same condition as that in Section 4.2.4, the following holds.

Neck growth

$$\frac{dV}{dt} = \frac{2\pi x^3}{a}\frac{dx}{dt} = \frac{D_l}{RT}\nabla\sigma\, A\, V_m$$

$$\approx \frac{D_l}{RT}\left(\frac{\gamma_s}{r}\frac{1}{r}\right)2\pi\frac{x^3}{a}V_m$$

$$\therefore x^5 = \frac{20 D_l \gamma_s V_m a^2}{RT} t \tag{4.21}$$

4.2.6 Evaporation/Condensation

In this mechanism, atoms evaporate from the sphere surface and the evaporated atoms condense in the neck region. When the distance between the evaporation region and the condensation region is shorter than the mean free path of gas atoms, the evaporation/condensation mechanism is the major mechanism of gas phase transport. When the distance is much longer than the mean free path, gas diffusion rather than evaporation/condensation becomes the major mechanism unless the reaction of gas atoms at the interface is slower than gas diffusion. The mean free path λ of gas atoms is inversely proportional to the total gas pressure in the system because λ is expressed as $\lambda = (\sqrt{2}\pi d^2 n)^{-1}$, where n is the number of atoms per unit volume and d the atom diameter. In real systems where sintering occurs by gas phase transport mechanisms, material transport from the particle surface to the furnace wall must also occur. Gas phase transport may be considered as a material transport mechanism during sintering rather than as a sintering mechanism.

As the evaporation or condensation of material is basically controlled by reactions of atoms at the surface, the kinetics of sintering by the evaporation/ condensation mechanism is also determined by either evaporation or condensation of atoms. The kinetics of neck growth by this mechanism was derived[5] from the Langmuir equation, a gas adsorption equation. The use of the Langmuir equation means that the condensation of atoms controls the neck growth.

Neck growth

According to the Langmuir adsorption equation, the amount (weight) of material deposited per unit area and per unit time is expressed as

$$m = \alpha \Delta p \left(\frac{M}{2\pi RT}\right)^{1/2} \tag{4.22}$$

where α is the sticking coefficient and M the molar weight of the material. If the atom deposited does not evaporate again, α equals 1.

Therefore,

$$\frac{dx}{dt} = \frac{m}{d} = \left(p_\infty \frac{\gamma_s}{r} \frac{V_m}{RT}\right)\left(\frac{M}{2\pi RT}\right)^{1/2} \Big/ d \tag{4.23}$$

where $d(=M/V_m)$ is the material density. Then,

$$\therefore x^3 = \sqrt{\frac{18}{\pi}} \frac{p_\infty \gamma_s}{d^2}\left(\frac{M}{RT}\right)^{3/2} at \tag{4.24}$$

4.2.7 Gas Diffusion

When gas diffusion is slower than the interface reaction, the neck growth is controlled by the diffusion of gas atoms from the particle surface to the neck surface. If the concentration gradient of gas species is extended to a distance of the neck curvature radius from the neck surface, as in the case of surface diffusion (Section 4.2.4), the neck growth can be calculated as follows.

Neck growth

$$A\frac{\mathrm{d}x}{\mathrm{d}t} = A\,D_g\,\nabla C\,V_m$$

$$= AD_g\frac{\nabla p}{RT}V_m \approx AD_g\frac{\Delta p}{RTr}V_m$$

$$\therefore\ \frac{\mathrm{d}x}{\mathrm{d}t} = D_g\frac{V_m}{(RT)^2}\frac{\gamma_s}{r^2}p_\infty V_m$$

$$\therefore\ x^5 = 20p_\infty D_g\gamma_s\left(\frac{V_m}{RT}\right)^2 a^2 t \qquad\qquad (4.25)$$

where D_g is the diffusivity of gas atoms and p the vapour pressure of the solid. The gas diffusivity is expressed as $D_g = \lambda\bar{c}/3$, where λ is the mean free path of gas atoms and $\bar{c}$ the mean velocity. Since $\bar{c}$ is equal to $(8RT/\pi M)^{1/2}$, where M is the molar mass, D_g is inversely proportional to the total gas pressure in the system.

4.3 SINTERING DIAGRAMS

As explained in Section 4.1.2, differences in bulk capillary pressure, bulk vacancy concentration and vapour pressure appear simultaneously and independently according to a difference in the surface curvature of particles in contact. Therefore, simultaneous and independent material transport caused by these differences occurs during sintering. However, there is a dominant mechanism that depends on the system concerned, sintering conditions, degree of sintering, etc.

   Ashby developed a sintering diagram based on various sintering mechanisms, such as those in Table 4.1. Under the assumption of no grain growth, the Ashby sintering diagram identifies the dominant sintering mechanism under various experimental conditions and shows the rate of sintering that results from all the mechanisms acting together (for example, Figure 4.4).[27,28] (The no grain growth assumption, however, may not be justified in real sintering, in particular for intermediate and final stage.) In Figure 4.4, the boundary lines between regions of dominant mechanisms indicate the experimental conditions under which the contributions of two different mechanisms to sintering (neck growth) are the same. In the Ashby sintering diagram, sintering is divided into three stages: stage 0 where adhesion between particles occurs at the very beginning of sintering, stage 1 where the driving force for sintering decreases as the neck grows, and stages 2 and 3 where the driving force increases with neck

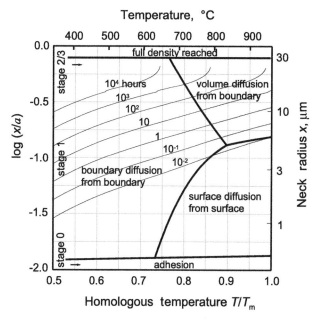

Figure 4.4. Sintering diagram of an aggregate of pure silver spheres of 38 μm radius.[27]

growth (spherical pore stage). The neck growth rate $(dx/dt)_t$ is expressed as $(dx/dt)_t = \sum_{i=1}^{6} (dx/dt)_i$ for stage 1 (for crystalline particles, the viscous flow mechanism is excluded) and $(dx/dt)_t = (dx/dt)_1 + (dx/dt)_2$ for stage 2, where i in $(dx/dt)_i$ represents the ith mechanism in Table 4.1. However, for a given experimental condition and neck size, the neck growth is dominated by one sintering mechanism, as shown in Figure 4.4. For example, when spherical silver particles with a radius of 38 μm are sintered at 0.8 T_m, the diagram shows that after particle adhesion at the very beginning, the neck growth occurs dominantly by surface diffusion, grain boundary diffusion and lattice diffusion in temporal sequence. The sintering diagram delineates not only the dominant mechanisms for various experimental conditions but also the sintering kinetics. The contours of constant time in Figure 4.4 show the neck sizes found after sintering for those periods of time at various temperatures.

4.4 EFFECT OF SINTERING VARIABLES ON SINTERING KINETICS

In general, the sintering rate (densification rate) increases with decreased particle size and with increased sintering temperature and time, as shown schematically in Figure 4.5. A quantitative explanation of this general tendency can be drawn from the kinetic equations derived in Section 4.2 and also summarized in Table 4.2.

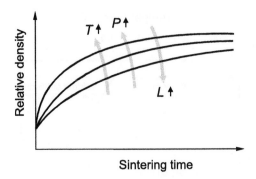

Figure 4.5. Effect of sintering parameters on densification.

Table 4.2. Summary of kinetic equations for various mechanisms of initial stage sintering

Sintering mechanism	Neck growth	Shrinkage	Scale exponent α
1. Lattice diffusion from grain boundary to neck	$x^4 = \dfrac{16 D_l \gamma_s V_m a}{RT} t$ $\equiv C_l D_l a t$	$\dfrac{\Delta l}{l} = \left(\dfrac{D_l \gamma_s V_m}{RTa^3}\right)^{1/2} t^{1/2}$	3
2. Grain boundary diffusion from grain boundary to neck	$x^6 = \dfrac{48 D_b \delta_b \gamma_s V_m a^2}{RT} t$ $\equiv C_b D_b \delta_b a^2 t$	$\dfrac{\Delta l}{l} = \left(\dfrac{3 D_b \delta_b \gamma_s V_m}{4RTa^4}\right)^{1/3} t^{1/3}$	4
3. Viscous flow	$x^2 = \dfrac{4 \gamma_s a}{\eta} t \equiv C_{vf} \dfrac{1}{\eta} a t$	$\dfrac{\Delta l}{l} = \dfrac{3 \gamma_s}{8 \eta a} t$	1
4. Surface diffusion from particle surface to neck	$x^7 = \dfrac{56 D_s \delta_s \gamma_s V_m a^3}{RT} t$ $\equiv C_s D_s \delta_s a^3 t$		4
5. Lattice diffusion from particle surface to neck	$x^5 = \dfrac{20 D_l \gamma_s V_m a^2}{RT} t$ $\equiv C_l' D_l a^2 t$		3
6. Gas phase transport			
6.1. Evaporation– condensation from particle surface to neck	$x^3 = \sqrt{\dfrac{18}{\pi}} \dfrac{p_\infty \gamma_s}{d^2} \left(\dfrac{M}{RT}\right)^{3/2} a t$ $\equiv C_{e/c} p_\infty a t$		2
6.2. Gas diffusion from particle surface to neck	$x^5 = 20 p_\infty D_g \gamma_s \left(\dfrac{V_m}{RT}\right)^2 a^2 t$ $\equiv C_g p_\infty D_g a^2 t$		3

4.4.1 Particle Size

The effect of particle size on sintering is well explained by Herring's scaling law.[29] When powders with similar shapes but different sizes are sintered under the same experimental conditions and by the same sintering mechanism, the scaling law predicts the relative periods of sintering time required to get the same degree of sintering. For the sintering of two kinds of powders with radii a_1 and a_2, where $a_2 = \lambda a_1$, the required sintering times t_2 and t_1 are interrelated as

$$t_2 = (\lambda)^\alpha t_1 \tag{4.26}$$

where α is an exponent. In the case of sintering which occurs by lattice diffusion, the sintering time t needed to get a given change in volume is expressed as

$$t = \frac{V}{JAV_m} = \frac{L^3}{((D_l/RT)(2\gamma_s/L)(1/L))L^2 V_m} \propto L^3 \tag{4.27}$$

where L is the length. Therefore, $\alpha = 3$ for lattice diffusion. This means that the period of time required to get the same degree of sintering for powders of different sizes is proportional to $(\lambda)^3$.

The exponent α in the scaling law can also be deduced from the sintering equations in Table 4.2 without following Herring's derivation. The sintering equations can be expressed as a general form

$$\left(\frac{x}{a}\right)^n = F(T) a^{m-n} t \tag{4.28}$$

To sinter powder compacts of different sizes to a given sintering degree (i.e. $x/a = $ constant), $a^{m-n}t$ must be constant and therefore $\alpha = n - m$. The value of α for each sintering mechanism is shown in the last column of Table 4.2.

The basic assumptions for the scaling law to be applicable are that the same sintering mechanism is maintained during the sintering of different powders and that the microstructure evolves with a similar shape. In reality, however, the similar shape assumption is generally not satisfied in sintering of powder compacts because the mechanism of grain growth is, in general, different to the mechanism of densification (see Section 11.5). Nevertheless, Herring's scaling law demonstrates in a simple way the effect of particle size on microstructural change in compacts being sintered under the same sintering mechanism.

4.4.2 Temperature

The effect of temperature can also be predicted from the equations in Table 4.2. Since sintering is a thermally activated process, the variables sensitive to temperature are diffusivity, viscosity, etc. which are expressed as exponential functions of temperature. Therefore $\ln t$ (natural log t) is at first proportional to

$1/T$, where t is the sintering time to get a given degree of sintering. However, the exact relationship is different for different mechanisms. In the case of lattice diffusion where $(x/a)^n \propto (D_l/T)a^{m-n}t$, $\ln(t/T)$ is proportional to $1/T$ with a slope of Q_l/R, where Q_l is the activation energy of lattice diffusion and R the gas constant. (See also Section 11.6.)

4.4.3 Pressure

The equations in Table 4.2 were derived for systems where the driving force of sintering is the capillary pressure difference due to curvature difference. However, when an external pressure $P_{appl.}$ is applied, the total sintering pressure P_t is the sum of the capillary and external pressures,

$$P_t = \frac{\gamma_s}{r} + P_{appl.} f(\rho, geo) \tag{4.29}$$

where $f(\rho, geo)$ is a function of relative density and particle geometry (see Section 5.6.1). Therefore, sintering equations for systems with external pressure are different to those without it. The densification rate, however, always increases with increased sintering pressure (Figure 4.5). Various techniques are available in pressure-assisted sintering, for example, gas pressure sintering, hot pressing and hot isostatic pressing. (See Section 5.6.)

4.4.4 Chemical Composition

In diffusion-controlled sintering, since atom diffusivity and, thus, atom mobility increases with an increased vacancy concentration, we can enhance sintering kinetics by increasing the vacancy concentration. For ionic compounds, vacancy concentration varies considerably with dopant addition. (See Section 13.1.)

4.5 USEFULNESS AND LIMITATIONS OF THE INITIAL STAGE SINTERING THEORY

All of the sintering equations were derived under the assumption that a local equilibrium of atoms with capillary pressure is maintained everywhere, in the atom source and in the atom sink. (This assumption is acceptable.) The dihedral angle of 180° is also an acceptable assumption because the dihedral angle affects only numerical constants in kinetic equations, and not the sintering variables (Eq. (4.7)).

Although the sintering equations were derived using the simple two-particle model, their usefulness resides in demonstrating the effects of sintering variables and physical parameters on sintering kinetics. In other words, they show what the processing variables in sintering are and what their effects will be. In addition, by comparing kinetic equations with each other, we can also evaluate relative contributions of sintering mechanisms under given experimental conditions. The evaluation of relative contributions is highlighted in Ashby's sintering diagram.[27,28]

In early studies on sintering, a number of investigations were made to experimentally determine sintering mechanisms by measuring neck growth and shrinkage kinetics. However, many of these investigations were misinterpreted because of inherent problems. The experimental conditions were sometimes in the ranges where two or three mechanisms were operative.[27] Therefore, the exponent n in sintering equations may have a range of values. Furthermore, the difference in the value of n among various mechanisms is not large, as shown in Table 4.2; it is improbable that the value of n will be exactly determined by experiments.

Therefore, care must be taken in discussing sintering mechanisms based on previous investigations. Some investigations determined diffusivities from sintering experiments. However, the diffusivities determined in these experiments may have no absolute meaning because of the inherent problems and may imply only that related diffusion mechanisms were operative in the experimental conditions studied.

5

INTERMEDIATE AND FINAL
STAGE SINTERING

When necks form between particles in real powder compacts, pores form interconnected channels along 3-grain edges. As the sintering proceeds, the pore channels are disconnected and isolated pores form because the dihedral angle is much larger than 60° and the shrinkage of the interconnected pores is not uniform due to the non-constant size of the pore channels and also Rayleigh's surface instability.[30,31] At the same time, the grains grow. Coble proposed two geometrically simple models for the shape changes of pores during intermediate and final stage sintering: the channel pore model and the isolated pore model, respectively.[32,33]

5.I INTERMEDIATE STAGE MODEL

Coble's microstructure model of intermediate stage sintering is based on bcc-packed tetrakaidecahedral grains with cylinder-shaped pores along all of the grain edges, as shown in Figure 5.1(a). This intermediate stage model assumes equal shrinkage of pores in a radial direction. Although the model is limited in terms of describing real sintering, it reasonably simplifies sintering complexity and allows the evaluation of the effect of sintering variables on sintering kinetics.

If the edge effect shown in the geometrical model (Figure 5.1(a)) is neglected, the atom flux towards the cylindrical pores may be similar to the thermal flux towards the pores provoked by an electrically heated wire with diameter equal to the grain boundary diameter. Let λ be the thermal conductivity, the thermal flux per unit length J_{heat} is $J_{heat} = -\lambda(dT/dx)$ and its solution $J_{heat} = 4\pi\lambda\Delta T$. Therefore, the atom flux per unit length J_{atom} is $J_{atom} = -(D/RT)(d\sigma/dx)$ and its solution $J_{atom} = 4\pi(D/RT)\Delta\sigma$. For the atom flux, two mechanisms are available: lattice diffusion and grain boundary diffusion.

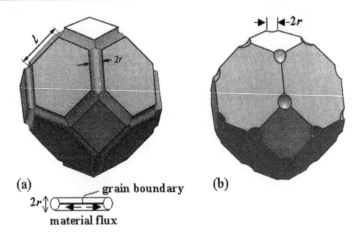

Figure 5.1. Coble's geometrical models for (a) intermediate stage and (b) final stage sintering.

5.1.1 Lattice Diffusion

Since pore shrinkage occurs at all of the 14 surfaces of the grain, the rate of pore volume change dV_p/dt is expressed as

$$\frac{dV_p}{dt} = \frac{-14}{2} 2r J_{\text{atom}} V_m$$
$$= -14 r 4\pi \frac{D_l}{RT} \left(\frac{\gamma_s}{r}\right) V_m \qquad (5.1)$$

Therefore, the rate of porosity change dP_v/dt is

$$\frac{dP_v}{dt} = \frac{dV_p}{dt} / 8\sqrt{2} l^3 = -\frac{d\rho}{dt} \qquad (5.2)$$

where l is the length of a grain edge and ρ is the relative density.

$$\therefore \frac{d\rho}{dt} = \frac{7\pi D_l \gamma_s V_m}{\sqrt{2} l^3 RT} = \frac{336 D_l \gamma_s V_m}{RTG^3} \qquad (5.3)$$

where G is the grain diameter which satisfies $(\pi/6)G^3 = 8\sqrt{2} l^3$. If grains do not grow during sintering (that is, when $G = \text{const.}$), the integration of Eq. (5.3) is simple. However, in reality, grain growth usually occurs. The grain size must then be expressed as a growth equation in order to integrate Eq. (5.3). (See Sections 11.4 and 11.5.)

5.1.2 Grain Boundary Diffusion

A similar procedure to that for lattice diffusion gives

$$\frac{dV_p}{dt} = -\frac{14}{2} 4\pi \frac{D_b}{RT} \frac{\gamma_s}{r} \delta_b V_m \tag{5.4}$$

where δ_b is the diffusion thickness of grain boundary diffusion. Therefore,

$$\frac{dP_v}{dt} = \frac{dV_p}{dt} / 8\sqrt{2} l^3 = -\frac{d\rho}{dt}$$

$$= -28\pi \frac{D_b \delta_b \gamma_s V_m}{RT} \frac{1}{8\sqrt{2} l^3} \frac{1}{\sqrt{P_v}} \sqrt{\frac{12\pi}{8\sqrt{2} l^2}}$$

$$= -854 \frac{D_b \delta_b \gamma_s V_m}{RTG^4} \left(\frac{1}{P_v}\right)^{1/2} \tag{5.5}$$

5.2 FINAL STAGE MODEL

As a geometrical model of final stage sintering, Coble took tetrakaidecahedral grains with spherical pores with a radius of r_1 at their corners, as shown in Figure 5.1(b). For this model, he suggested concentric sphere diffusion of atoms from a distance r_2 to the pore surface. The atom flux passing through any concentric sphere surface with a radius of r, J_{total}, is constant, i.e.

$$J_{total} = const. = -4\pi r^2 \frac{D_l}{RT} \frac{d\sigma}{dr} \tag{5.6}$$

Therefore, by integration,

$$J_{total} = 4\pi \frac{D_l}{RT} \Delta\sigma \frac{r_1 r_2}{r_2 - r_1} \tag{5.7}$$

If $r_1 \ll r_2$,

$$\frac{d\rho}{dt} = -\frac{24}{4} J_{total} V_m / \frac{1}{6} \pi G^3$$

$$= \frac{288 D_l \gamma_s V_m}{RTG^3} \tag{5.8}$$

This equation indicates that the densification rate is inversely proportional to the cube of grain size. This result is the same as that found for the dependence of neck growth and shrinkage on particle size in the initial stage model.

So far, Coble's model has been a standard for interpreting and predicting the densification at final stage sintering. In Coble's model, however, a fundamental aspect is not taken into account, namely, the grain boundary as the atom source for densification. In addition, it is hard to accept Coble's flux equation (Eq. (5.7) under $r_1 \ll r_2$) that predicts a constant material flux from grain boundaries to the surface of a pore irrespective of pore size.

Unlike Coble's concentric sphere diffusion model, Herring's scaling law concept can be utilized to predict the sintering kinetics at the final stage.[34,35] This concept accounts for the effect of the surface area of the pore on the material flux from the grain boundary to the pore.[35]* Adopting this concept, Kang and Jung[35] derived the densification rate not only for volume diffusion but also for grain boundary diffusion. Since a stress gradient is thought to exist from the pore surface to the centre of the grain boundary, as demonstrated by a previous calculation,[17] the gradient may be assumed to be present over a distance of $l/2$. Then, using the relationship $8\sqrt{2}l^3 = (\pi/6)G^3$,

$$\frac{d\rho}{dt} = \frac{441 D_l \gamma_s V_m}{RTG^3}(1 - \rho)^{1/3} \qquad (5.9)$$

is obtained for lattice diffusion, and

$$\frac{d\rho}{dt} = \frac{735 D_b \delta_b \gamma_s V_m}{RTG^4} \qquad (5.10)$$

for grain boundary diffusion. In the case of grain boundary diffusion the dependence of densification rate on grain size in Eq. (5.10) is the same as that in the initial stage model. For lattice diffusion, the size dependence in Eq. (5.9) is the same as that in Eq. (5.8) of Coble's model. However, Eq. (5.9) contains a relative density term.

At the final stage the densification is interrelated with grain growth in the presence of pores, as described in Chapter 11. Figure 5.2(a) shows the calculated densification curves at 1727°C of Al$_2$O$_3$ powder compacts with powders 0.8 and 4.0 μm in size from Eqs (5.9) and (5.10) for densification and Eq. (11.20) with a numerical constant of 110 for grain growth.[36] (Here, grain growth is assumed to be governed by surface diffusion. When another mechanism governs the grain growth, the corresponding equation must be used for the calculation. The grain growth mechanism may also change during densification. As the pore size considerably decreases with densification,

---

*This concept can also be applied to the intermediate stage of sintering. Coble's intermediate stage model also took the surface area of the pore into account. The kinetic equations based on the two models are, in fact, the same except for the numerical constants.

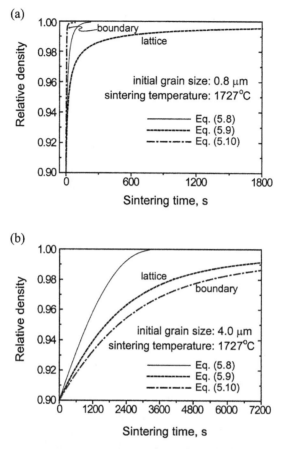

Figure 5.2. Calculated densification curves at final stage sintering at 1727°C of an alumina powder compact with (a) 0.8 and (b) 4.0 μm grain size at 90% relative density. Curves calculated using Coble's equation (Eq. (5.8)) are also presented as a comparison.[35]

not the pore migration but the boundary mobility itself should govern the grain growth (boundary control). (See Sections 11.1–11.3.) For such a very final stage, Eq. (11.20) must be replaced with an appropriate equation to give a better prediction.) The calculated curves show that densification rate decreases with increasing particle size and also increasing sintering time for the same powder, in agreement with experimental observations. In terms of densification mechanisms, however, the contribution of grain boundary diffusion dominates that of lattice diffusion for the fine powder (0.8 μm) and vice versa for the coarse powder. This result is consistent with the scale effect on sintering mechanisms; reduction in scale enhances the relative contribution of grain boundary and surface diffusion over lattice diffusion.

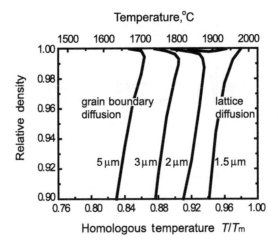

Figure 5.3. A sintering diagram of alumina at final stage sintering. Fields of the dominant mechanism, either grain boundary diffusion or lattice diffusion, are shown for various grain sizes at the beginning of the calculation (90% relative density).[35]

Figure 5.3 plots a final stage sintering diagram of alumina with various grain sizes at the beginning of the calculation (90% relative density). This figure shows that for the usual temperature range of sintering of commercial Al_2O_3 powders, 0.7–0.8 of the homologous temperature, the densification occurs by grain boundary diffusion. This conclusion should generally be correct for final stage sintering of any kind of commercial powder. At high temperatures lattice diffusion can dominate the densification. However, as sintering proceeds, the dominant mechanism can change to grain boundary diffusion as the curves in Figure 5.3 show. This result is due to the reduction in pore size with densification. As the pore size reduces considerably, grain growth can be much enhanced (see Section 11.2). When pores are sufficiently mobile to migrate along with the boundaries and are not entrapped within grains, the number density of pores per grain is not expected to change considerably.[37] A steady state should be maintained. As the grain growth rate increases quickly with densification, under certain conditions, pore size can increase during extended sintering for final densification although sintered density increases with sintering time. A series of microstructures of dry H_2-sintered Al_2O_3 in a previous investigation[37] supports this expectation. The change in dominant mechanism from grain boundary diffusion to lattice diffusion at the very final stage of sintering (over ~99% relative density in Figure 5.3) is due to the increase in pore size that results from pore coalescence with grain growth. However, when grain growth is not fast and hence pore growth does not occur, the change in sintering mechanism from grain boundary diffusion to lattice diffusion does not occur and the final densification is always governed by grain boundary diffusion.

5.3 ENTRAPPED GASES AND DENSIFICATION

The final densification of powder compacts is strongly affected by the sintering atmosphere because the atmospheric gas is entrapped within pores as they are isolated. In the case of a fast diffusing gas, full densification is possible, but this is impossible in a slowly diffusing or inert gas unless a high external pressure is applied.[38,39]

The effect of entrapped inert gases on densification can be calculated for various experimental conditions, as shown by Kang and Yoon.[40,41] They calculated the final density and densification kinetics as functions of initial pore radius r_i, surface energy γ_s, pore coalescence, dihedral angle, etc. Figure 5.4[40] shows the change in driving pressure of pore shrinkage when inert gases are entrapped within the pore. Figure 5.4(a) illustrates the case just before pore isolation, Figure 5.4(b) the shrinkage stage of the isolated pore and Figure 5.4(c) the final state where no further shrinkage occurs. When the pressure of entrapped gases becomes equal to the capillary pressure of the pore, the pore stops shrinking and the maximum attainable sintered density is reached (Figure 5.4(c)).

If the entrapped gas behaves like an ideal gas,

$$P_i\left[\left(r_i/r_f\right)^3-1\right] = 2\gamma_s/r_f \tag{5.11}$$

where r_f is the final radius of the isolated pore. Using Eq. (5.11), the maximum attainable densities for various initial pore sizes can be calculated, as shown in Figure 5.5.[40] For the calculation, the isolation of mono-size spherical pores was assumed to occur at 93% of the relative density. However, Figure 5.5 can also be used to predict the maximum attainable densities for pore isolation at any relative density, say $(1-\alpha)\%$, as shown on the right ordinate in the figure. According to this calculation, if the pore radius is less than a few microns,

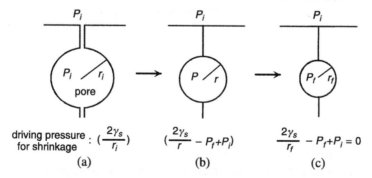

Figure 5.4. Schematic of pore shrinkage during sintering:[40] (a) just before the isolation of the pore, (b) shrinkage stage, and (c) final state.

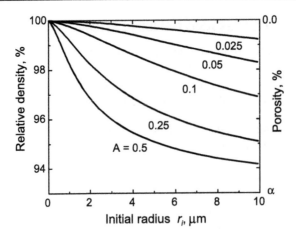

Figure 5.5. Maximum attainable density versus r_i, with an initial porosity of $\alpha\%$ $(A = P_i/2\gamma_s\ \mu m^{-1})$.[40] The ordinate is scaled by assuming an initial density of 93% of the theoretical value.

the maximum attainable density is over 99.8% in conventional sintering at 1 atm pressure and entrapped inert gases have essentially no effect. On the other hand, when the initial gas pressure is high or initial pore size is large, the maximum attainable density decreases considerably. Although Figure 5.5 was obtained for powder compacts without pore coalescence during sintering, it can also be utilized to predict the effect of pore coalescence. When pore coalescence occurs, reduction in sintered density (dedensification) results. In the case of coalescence of n pores, scaling of the abscissa by $r_i\ n^{1/3}$ should permit an estimate of the coalescence effect.

Application of an external pressure after pore isolation increases sintered density. Figure 5.6[41] shows the calculated maximum attainable density of powder compacts with external pressure, B atm, after pore isolation in an inert atmosphere of 1 atm, for example, in two-stage sintering or sinter-HIP (hot isostatic pressing). Regardless of initial pore size and up to a few tens of microns, essentially full densification is obtained by applying a pressure of an order of 100 atm. This result suggests that applying a pressure of a few thousand atm, as in hot isostatic pressing, has no advantage in improving sintered density. Instead, the sintering time for full densification is reduced because the densification rate increases and, in addition, the densification mechanism might also change with an increased external pressure. (See Section 5.6.) In the case of lattice diffusion, as the densification rate is linearly proportional to sintering pressure, the densification time is inversely proportional to sintering pressure. The effect of pore coalescence in pressure sintering (Figure 5.6) can also be predicted to be similar to that in pressureless sintering in a 1 atm gas atmosphere (Figure 5.5).

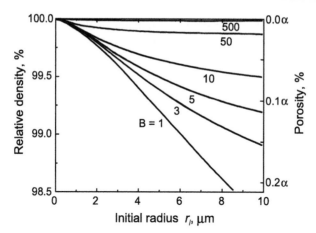

Figure 5.6. Maximum attainable density versus initial pore radius, r_i, with an initial porosity of $\alpha\%$, for various external pressures, B atm, after the isolation of pores which contain insoluble gases at I atm.[41] The ordinate is scaled by assuming an initial density of 93% of the theoretical value.

5.4 SINTERING PRESSURE AT FINAL STAGE SINTERING

The sintering pressure at the final stage is usually considered to be the capillary pressure due to pore curvature, i.e. $2\gamma_s/r$. However, by using the thermo-dynamic concept that sintering pressure is the change in total free energy with respect to total volume change during densification (similar to the concept of effective pressure in a two-phase system in equilibrium (see Section 3.3)), Raj[42] calculated sintering pressure P_0 to be

$$P_0 = \frac{2\gamma_b}{G} + \frac{2\gamma_s}{r} \tag{5.12}$$

where γ_b is the grain boundary energy and G the grain size. The derivation was made under the following assumptions:

(i) the ratio of the number and the type of pores per grain remain constant,
(ii) a quasi-equilibrium shape of pores is maintained, and
(iii) the grains are nearly spherical or equiaxed.

The equation appears to be thermodynamically correct. However, it is different to the capillary pressure equation

$$\Delta P = \frac{2\gamma_s}{r} \tag{5.13}$$

At present, it seems that ambiguity still remains in clarifying the real sintering pressure.[43,44] However, since $\gamma_s > \gamma_b$ and $G \gg r$ at the final stage, we may safely use Eq. (5.13) as the sintering pressure.

5.5 POWDER PACKING AND DENSIFICATION

Most sintering models deal with ideal systems where particles of a constant size are uniformly packed and thus pores of the same size are uniformly distributed. Densification and shrinkage of powder compacts are, therefore, assumed to occur uniformly throughout the compacts. However, in real powder compacts, the sizes of the particles are variable, and pore size and distribution are non-uniform. Because of the non-uniform distributions of particles and pores of non-uniform sizes, differential sintering and densification occur and particles may move, which can cause the generation of pores larger than the initial pores.[45,46] Grain growth at the initial stage of sintering can also cause the formation of large pores.[47,48] Therefore, as the sintering proceeds, large pores can be locally generated, and the phenomenon of pore opening is observed. This means that the particle packing and distribution are an important parameter which determines the sintering kinetics of real powder compacts. In an experiment using Cu particles on a substrate, Petzow and Exner[45] experimentally demonstrated and theoretically analysed the particle rearrangement that may occur during solid state sintering. According to their results, asymmetric arrangement of the particles surrounding the neck causes particle rearrangement. When the neck geometry is asymmetric, the material flux from grain boundary to neck is non-uniform and more material comes to the neck region with a smaller radius of curvature. The result of this non-uniform material flux is an inclination of particles towards the direction of the smaller curvature radius.

In a model experiment using glass spheres, Liniger and Raj[46] showed the effect of two-dimensional packing of particles with two different sizes. When the glass spheres were mono-sized, the degree of hexagonal close packing was high but large defects remained between close-packed domains. The large defects became larger as the sintering proceeded. In contrast, when spheres of two different sizes were mixed and sintered, the degree of hexagonal close packing was low, but large defects did not form and densification occurred uniformly. This result indicates that a mono-sized powder is not beneficial for densification and powder with an appropriate size distribution is more desirable. Therefore, in real powder processing, powders with a narrow size distribution should be used to suppress the formation of large flaws.

Large flaws may also form because of non-uniform compact density. When powder compacts are made using agglomerates of fine powders, the agglomerates may either disintegrate (soft agglomerates) or remain unchanged (hard agglomerates) during pressing. If the agglomerates are not fully disintegrated, the compact density is locally non-uniform and full densification is hard to

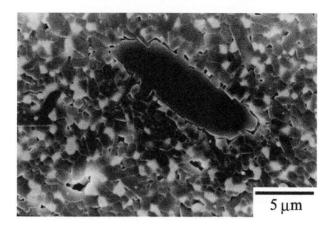

Figure 5.7. Tangential crack formed around an Al_2O_3 platelet in 80Mullite-10ZrO$_2$(TZ-3Y)-10Al$_2$O$_3$ (platelet) sintered at 1700°C for 1 h in air.[53] The dark and large area is an Al_2O_3 platelet, dark and small grains mullite grains, and bright and small grains ZrO$_2$ grains.

achieve because of differential densification during sintering.[49–51] A similar phenomenon can occur when using dissimilar powders, whiskers or platelets.[47,48] (See Section 5.7.)

When differential densification occurs, back stresses build up in the compact and large flaws can form.[49–52] Figure 5.7[53] shows, as an example, a tangential crack formed around a platelet. Above all, whenever a large defect forms during powder processing, the defect is apt to enlarge during sintering. Therefore, the formation of large defects during powder processing must be avoided in order to achieve full densification.

5.6 PRESSURE-ASSISTED SINTERING

Application of an external pressure to a powder compact results in a direct increase in the driving force of densification and an increase in densification kinetics. On the other hand, grain growth is not related to the applied external pressure. Therefore, the effect of an external pressure is more effective in systems where grain growth rate relative to densification rate is high. As the densification rate is enhanced by external pressure, the sintering temperature as well as the sintering time can be reduced and grain growth further suppressed.

Two different types of pressure application are available: unidirectional and isostatic. The former is typified by hot pressing (HP) and the latter by hot isostatic pressing (HIP). Hot pressing is used to unidirectionally press a powder compact in a die at a high temperature using a moderate pressure, 20–50 MPa for a graphite die. Hot isostatic pressing of powder compacts is usually done after encapsulation of the compacts within a container, using an inert gas of a higher pressure than that of HP, up to ~300 MPa. As a variation of HIP, gas pressure sintering (GPS) usually under 10 MPa is available. GPS makes possible the sintering of materials with a high vapour pressure, such as Si_3N_4. In GPS, two-stage sintering is usually performed: low pressure sintering before the isolation of pores and high pressure sintering after the pore isolation.[54]

In addition to sintering, HIP is also used in attempts to achieve full densification or to cure flaws remaining after sintering and this application is, in fact, more general than its role in pure sintering. In this case, since the pores are isolated in the sintered compacts, encapsulation is usually not necessary. If a high pressure is applied to a compact containing interconnected pores, on the other hand, HIP cannot completely densify the compact, as shown in Figure 5.8,[55] because the high pressure gas in the vessel is entrapped within the pores at the moment of pore isolation. Full densification by HIP can be achieved only after pore isolation with zero apparent porosity (interconnected porosity). On the other hand, when isolated pores are entrapped within grains due to pore boundary separation, full densification is also impossible by HIP as shown by the samples sintered at and above 1450°C in Figure 5.8.

Recently, a new HP technique where a pulsed electric field is applied has been developed, so-called spark plasma sintering (SPS) or pulsed electric current sintering (PECS).[56] The experimental set-up is similar to that of HP with a graphite die and graphite punches. The heating of the sample, however, is achieved by resistance heating caused by a pulsed current (typically a few thousand amperes) under a pulsed DC voltage (typically a few volts). A high heating rate of more than a few hundred K/min can be achieved. In SPS the densification is very fast, taking a few minutes, while grain growth is very limited at process temperatures which are measured to be lower than HP temperatures. The SPS technique has been successfully applied to a number of systems, for example Al_2O_3, ZrO_2, Si_3N_4 and MgB_2. However, the mechanism of the SPS process is unclear as yet.

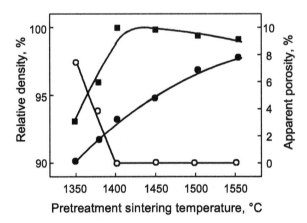

Figure 5.8. Effect of sintering temperature on the relative density (●) and the apparent porosity (○) before hot isostatic pressing, and on the relative density (■) after hot isostatic pressing.[55]

5.6.1 Driving Force of Pressure-Assisted Sintering

The driving force of densification under an external pressure is determined by the pressure itself and also the contact area relative to the cross-sectional area of the particles. At the initial stage of HP of a compact with a simple cubic packing of spherical particles, the effective compressive pressure P_1^* at the contact area is expressed as[57]

$$P_1^* \approx \frac{4a^2}{\pi x^2} P_{appl.} + \frac{\gamma_s}{r} \tag{5.14}$$

Here, a is the particle radius, x the neck radius and r the radius of curvature of the neck surface (Figure 4.2(b)). At the final stage of sintering, on the other hand, the effective pressure P_2^* is expressed as[57]

$$P_2^* \approx \frac{P_{appl.}}{\rho} + \frac{2\gamma_s}{r} \tag{5.15}$$

if the pores are uniformly distributed in the compact. Here, ρ is the relative density of the compact.

A similar consideration of geometry can be used to estimate the effective pressure under HIP. Provided that a force equilibrium is maintained at the surface of particles, the effective pressure P_1^* at the initial stage can be expressed as[57]

$$P_1^* = \frac{4\pi a^2}{\pi x^2 Z} P_{appl.} + \frac{\gamma_s}{r} \tag{5.16}$$

where Z is the number of neighbouring grains. Assuming that the particles are randomly packed, Arzt et al.[58] derived the effective pressures for compacts with $\rho < 0.9$ and $\rho > 0.9$ to be

$$P_1^* = \frac{4\pi a^2}{\pi x^2 Z \rho} P_{appl.} + \frac{\gamma_s}{r} \tag{5.17}$$

and

$$P_2^* \approx P_{appl.} + \frac{2\gamma_s}{r} \tag{5.18}$$

respectively.[†]

[†]These equations apply to the case with no entrapped gas and $x \gg r$.

5.6.2 Sintering Mechanisms in Pressure-Assisted Sintering

Several densification mechanisms can be operative in HP or HIP, as in pressureless sintering.[†] In addition to lattice and grain boundary diffusion, plastic deformation and creep, which are unimportant in pressureless sintering, can be major mechanisms. The major densification mechanism for a given system can vary with experimental and compact conditions, such as temperature, pressure, particle size and neck size.[58,59] However, the overall densification rate of a compact is, of course, the sum of the densification rates of all the mechanisms operative.

Figure 5.9[58] shows an example of HIP diagrams which identify the dominant densification mechanism under various experimental conditions and shows the rate of densification that results from all the mechanisms acting together. As can be seen in the diagrams, diffusion is usually the dominant mechanism in ceramics even under a high external pressure while power-law creep can be an important densification mechanism in metals.

Plastic deformation

At the early stage of densification where P_1^* is high, plastic yielding can be the major densification mechanism. The plastic deformation between particles may be regarded as identical to that occurring in a hardness test using an indenter. Then, if the indentation stress σ_i satisfies the yielding condition,

$$\sigma_i \approx 3\sigma_Y \tag{5.19}$$

densification by plastic deformation occurs. Here σ_Y is the yield stress of the material. As the contact area increases with plastic deformation and P_1^* becomes smaller than σ_i, plastic deformation stops. If the external hydrostatic pressure is so high that plastic deformation occurs even at the final stage of densification ($\rho \geq 0.9$), the densification may be simplified as an instantaneous plastic deformation of thick spherical shells. In this case the deformation condition can be expressed as[59]

$$P_2^* \geq \frac{2}{3}\sigma_Y \ln\left(\frac{1}{1-\rho}\right) \quad (\rho > 0.9) \tag{5.20}$$

Power-law creep

Power-law creep can also be a major densification mechanism in pressure-assisted sintering. At the early stage of densification achieved by power-law

---

[†]The terminology 'pressureless sintering' is used for atmospheric pressure sintering without an external pressure.

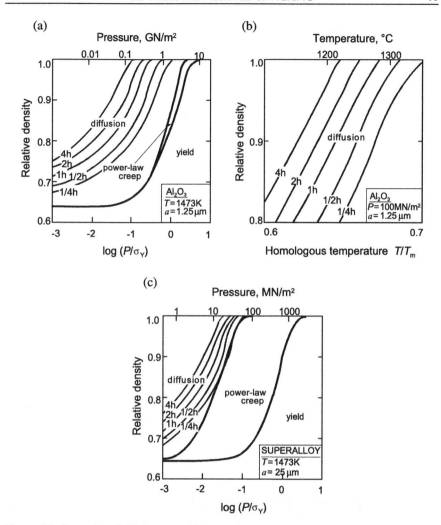

Figure 5.9. Examples of HIP diagrams: (a) density/pressure map $T = 1473$ K for alumina with a particle diameter of 2.5 μm, (b) density/temperature map at $P_{appl.} = 100$ MN/m^2 for alumina with a particle diameter of 2.5 μm and, (c) density/pressure map at $T = 1473$K for a superalloy with a particle diameter of 50 μm.[58]

creep, the densification rate $\mathrm{d}\rho/\mathrm{d}t$ can be expressed as

$$\frac{\mathrm{d}\rho}{\mathrm{d}t} = f(\rho, \mathrm{geo}) \frac{x}{a} \dot{\varepsilon}_o \left(\frac{P_1^*}{3\sigma_o} \right)^n \qquad (5.21)$$

if the creep is considered to be similar to that occurring during indentation of a spherical indenter.[58–60] Here, $\dot{\varepsilon}_o$, σ_o and n are material properties, and $f(\rho, \mathrm{geo})$

a function of the initial and sintering densities of the compact, and of particle geometry. The constant 1/3 in parentheses in the right-hand side of Eq. (5.21) implies that the geometrical arrangement of particles is similar to that for Eq. (5.19). When the particles are randomly packed, P_1^* is expressed as Eq. (5.17). The value of n in Eq. (5.21) is usually in the range between 3 and 8.[58,60]

For final stage densification by power-law creep under HIP, where a hollow sphere model (an isolated spherical pore within a spherical particle) is acceptable, $d\rho/dt$ can be expressed as[58–60]

$$\frac{d\rho}{dt} = f(\rho)\dot{\varepsilon}_o \left(\frac{3}{2n} \frac{P_2^*}{\sigma_o} \right)^n \tag{5.22}$$

Here, $f(\rho)$ is a complicated function of ρ and P_2^* is expressed as Eq. (5.18).

Diffusion

The densification by plastic deformation and power-law creep is, in principle, independent of particle (grain) size. In the case of diffusion (both lattice and grain boundary), on the other hand, densification depends on not only the effective pressure but also the grain size. The densification by diffusion under an external pressure is similar to diffusional creep: Nabarro–Herring creep due to lattice diffusion,[24,25] and Coble creep due to grain boundary diffusion.[26,57] The dependency of densification on grain size is the same as that of diffusional creep.

The densification rate by diffusion may be estimated following the derivation of a scaling law (see Section 4.4.1). In the case of lattice diffusion, for example, the time needed to get the same degree of densification, t, is expressed as

$$t = \frac{V}{JAV_m} = \frac{L^3}{(D_l/RT)(1/L)(P^*)L^2 V_m} \tag{5.23}$$

Therefore, the densification rate $d\rho/dt$ is expressed as

$$\frac{d\rho}{dt} \propto \frac{1}{t} \propto \frac{D_l V_m P^*}{RTa^2} \tag{5.24}$$

where a is the grain radius and P^* the effective pressure expressed by Eqs (5.17) and (5.18). Similarly, for grain boundary diffusion

$$\frac{d\rho}{dt} \propto \frac{D_b \delta_b V_m P^*}{RTa^3} \tag{5.25}$$

The exponent of grain size in Eqs (5.24) and (5.25) (the exponent α in a scaling law) is smaller by 1 than that in pressureless sintering: α equals 2 and 3 for lattice and grain boundary diffusion, respectively, instead of 3 and 4 in pressureless sintering. When the contributions of the two diffusion mechanisms are both considerable, the densification rate is expressed as the sum of Eq. (5.24) and Eq. (5.25)[58–60], i.e.

$$\frac{d\rho}{dt} = f(\rho, geo)\frac{(D_b\delta_b + rD_l)V_m}{RTa^3}P^*$$
(5.26)

Here, $f(\rho, geo)$ is a function of the relative compact density and grain geometry, and r the radius of curvature of the pore.

However, as noted by Kang and Jung,[35] the material transported from the grain boundary to a pore should be affected also by the surface area of the pore during densification (see Section 5.2). Following the same procedure of Kang and Jung[35] it is possible to calculate the densification rate at final stage sintering due to a HIP pressure for the geometry shown in Figure 5.1(b). Assuming that a stress gradient is present across the grain boundary, as in the case for Eqs (5.9) and (5.10), and that the effective pressure at the grain boundary is $P_{appl.}$ (Eq. (5.18)),

$$\frac{d\rho}{dt} = \frac{61D_lV_mP_{appl.}}{RTG^2}(1-\rho)^{2/3}$$
(5.27)

and

$$\frac{d\rho}{dt} = \frac{101D_b\delta_b V_mP_{appl.}}{RTG^3}(1-\rho)^{1/3}$$
(5.28)

are obtained for lattice and grain boundary diffusion, respectively. The dependence of densification rate on grain size in these equations is the same as that in Eqs (5.24) and (5.25). However, Eqs (5.27) and (5.28) contain relative density terms.

Comparison of Eqs (5.27) and (5.28) with Eqs (5.9) and (5.10) shows that the densification rate by the capillary pressure is higher than that by the externally applied pressure under certain conditions. This result is due to the reduction of pore size and the increase in capillary pressure above the external pressure with pore size reduction. The very final densification is thus expected always to be governed by pressureless sintering kinetics. The densification rate in pressure-assisted sintering is, of course, the sum of those due to the capillary pressure (Eqs (5.9) and (5.10)) and to the external pressure (Eqs (5.27) and (5.28)). The contribution of capillary pressure increases as the external pressure decreases and also as the pore size decreases with densification.

5.7 CONSTRAINED SINTERING

When differential sintering occurs, the sintering is constrained within boundary conditions. Typical examples are the sintering of composites with rigid inclusions, sintering of thin films on a substrate and co-sintering of different laminates. In thin film sintering on a rigid substrate a lateral constraint is imposed and only shrinkage perpendicular to the film is allowed. In the case of co-sintering of porous multi-layers an additional possibility of warping of the layers is introduced.

Constrained sintering can be well described macroscopically by using continuum mechanics.[61-63] If a powder compact is elastically isotropic and linearly viscous, the constitutive equations are written as

$$s_{ij} = 2G_p \dot{e}_{ij} \tag{5.29}$$

and

$$\sigma = K_p(\dot{\varepsilon} - 3\dot{\varepsilon}_f) \tag{5.30}$$

where s_{ij} ($i, j = x, y, z$) is the shear stress, G_p the shear viscosity, $\dot{e}_{ij}$ the shear strain rate ($= \dot{\varepsilon}_{ij} - (1/3)\delta_{ij}\dot{\varepsilon}$ [$\dot{\varepsilon}_{ij}$ the strain rate, δ_{ij} the Kronecker delta, $\dot{\varepsilon}$ the volumetric strain rate]), σ the mean (hydrostatic) stress, K_p the bulk viscosity, and $\dot{\varepsilon}_f$ the free strain rate. From the concept of linear elasticity

$$K_p = \frac{E_p}{3(1 - 2v_p)} \tag{5.31}$$

and

$$G_p = \frac{E_p}{2(1 + v_p)} \tag{5.32}$$

where E_p is the uniaxial viscosity and v_p the viscous Poisson's ratio. Note that E_p is a function of density and grain size while v_p is a function of density.

From Eqs (5.29)–(5.32)

$$\dot{\varepsilon}_x = \dot{\varepsilon}_f + E_p^{-1}[\sigma_x - v_p(\sigma_y + \sigma_z)] \tag{5.33a}$$

$$\dot{\varepsilon}_y = \dot{\varepsilon}_f + E_p^{-1}[\sigma_y - v_p(\sigma_x + \sigma_z)] \tag{5.33b}$$

and

$$\dot{\varepsilon}_z = \dot{\varepsilon}_f + E_p^{-1}[\sigma_z - v_p(\sigma_x + \sigma_y)] \tag{5.33c}$$

where $\dot{\varepsilon}_i (i = x, y, z)$ is the uniaxial strain rate in the i direction and σ_i the uniaxial stress in the i direction. For free sintering, where the free strain rate $\dot{\varepsilon}_f$ is governed by the hydrostatic sintering stress σ_s and the bulk viscosity K_p,

$$3\dot{\varepsilon}_f = \frac{\sigma_s}{K_p} = -\left(\frac{\dot{\rho}}{\rho}\right)_f \tag{5.34}$$

is satisfied. In this case the densification rate $(-\dot{\rho}/\rho)_f$ is directly correlated to the free strain rate. When normal stresses are applied onto the sintering body, the densification rate is expressed as

$$-\frac{\dot{\rho}}{\rho} = \dot{\varepsilon}_x + \dot{\varepsilon}_y + \dot{\varepsilon}_z$$
$$= \frac{\sigma_s + 1/3(\sigma_x + \sigma_y + \sigma_z)}{K_p} \tag{5.35}$$

from Eqs (5.33) and (5.34).

The constitutive parameters for a continuum mechanical description of sintering can be determined by measuring radial and axial densification rates while a well-defined uniaxial compressive stress is applied.[64-66] Alternatively, shrinkage can be stopped in one direction, if a uniaxial tensile stress (e.g. in a thin film) is applied, while shrinkage continues in the other directions.[67] This uniaxial tensile stress is a measure of the uniaxial sintering stress. In predicting sintering stresses and sintering viscosities, surface diffusion is typically assumed to be fast so that equilibrium surface geometries are maintained at the intermediate and the final stage of sintering.[44,68,69] The sintering stresses and viscosities then depend only on grain size, relative density and the initial grain assembly.[44,69] Some models and experimental results are available on this subject.[44,65,70-72]

In the case of thin film sintering the lateral constraint imposed by a rigid substrate allows a shrinkage only in the direction (z-direction) perpendicular to the film.[73-75] The imposed constraint induces an in-plane tensile stress and the stress accelerates the shrinkage in the z-direction. From the lateral constraint $\dot{\varepsilon}_x = \dot{\varepsilon}_y = 0$, $\sigma_z = 0$, and $\sigma_x = \sigma_y = \sigma$, and the expressions of the uniaxial strain rates in Eq. (5.33), the densification rate in the z-direction is expressed as

$$\dot{\varepsilon}_z = \left(\frac{1 + \nu_p}{1 - \nu_p}\right)\dot{\varepsilon}_f \tag{5.36}$$

This equation correlates the densification rate of a constrained film, $\dot{\varepsilon}_z$, with the densification rate of a free sintered body, $\dot{\varepsilon}_f$, via the viscous Poisson's ratio, ν_p. As the grain size is comparable to the thickness of a thin film, however, film thickness affects the viscosity and the sintering stress of the film.[76] Densification as well as grain growth can be suppressed considerably.[76] When several porous and sinter-active layers are fired together (co-fired), both the densification and the warpage of the layers are of importance. The densification and warpage can also be predicted well by use of constitutive sintering parameters that can be evaluated by sintering monolithic materials.[77] However, when a sacrificial layer, which serves as a rigid substrate but needs to be removed for the application at hand, is used to meet stringent dimensional requirements, co-sintering then becomes a variant of thin film sintering with several layers sintered on top of a sacrificial rigid substrate.

PROBLEMS

2.1. With a decreased particle size, what sintering mechanisms become relatively more important? Why?

2.2. Consider two glass spheres in contact just below their melting point. Draw schematically their shape change and describe its kinetics with annealing time. No gravity effect is assumed.

2.3. Consider that two inert gas-containing spherical pores of the same size come into contact within a single crystal, as shown in Figure P2.3. Describe the microstructural changes and their processes during annealing of this crystal. Assume that the pore volume does not change during annealing and that the equilibrium shape of the pores is a sphere.

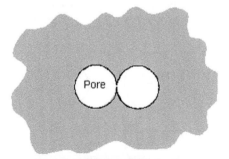

Figure P2.3

2.4. (a) Explain Herring's scaling law in solid state sintering and describe important points to be considered in its application to real systems.

(b) Derive the scaling law for the evaporation/condensation mechanism.

2.5. Which process, above all, controls sintering achieved by the evaporation/condensation mechanism?

2.6. Derive the scaling law of sintering by lattice diffusion. Under an applied external pressure ($P_{appl.} \gg 2\gamma/r$), how should the scaling law be modified?

2.7. Draw schematically and explain the changes in the driving force of densification with relative density of

(i) close-packed monosize particles and

(ii) a powder compact with a particle size distribution and hence grain growth during sintering.

2.8. Describe the shape of two different sized particles in contact during sintering. What is the driving pressure of sintering of the particles by grain boundary diffusion?

2.9. Consider a two-particle system where the neck growth occurs through gas phase transport. Explain the dependence of sintering time on temperature for
  (i) gas diffusion and
  (ii) evaporation/condensation.
  Assume that the gas pressure in the system is the vapour pressure of the material at the temperature concerned.

2.10. Consider a system where the sintering occurs by gas phase transport. Explain in detail the change in neck growth rate (dx/dt) with sintering temperature by the evaporation/condensation mechanism in vacuum sintering. Draw schematically and explain the change in neck radius with external Ar gas pressure (from zero to several thousand atm) for the same temperature and the same period of time ($\log x$ versus $\log P_{Ar}$).

2.11. Consider three crystalline particles separated, as shown in Figure P2.11. Draw schematically and explain the change in their shape with annealing time if material transport occurs via a gas phase. Assume that all of the material evaporated is transported between particles and that the distance between particles A and B is very small.

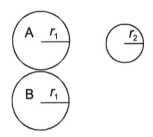

Figure P2.11

2.12. In the two-particle model of initial stage sintering, the neck growth is expressed as $(x/a)^n = F(T)a^{m-n}t$ (Eq. (4.28)), where x is the neck radius and a the particle radius. The equation is, in general, acceptable for $(x/a) < 0.2$. With increasing values of (x/a) above 0.2, would the exponent n become smaller or larger? Explain.

2.13. In sintering by grain boundary diffusion, material redistribution by either lattice or surface diffusion is required. Calculate roughly the control limit of surface diffusion for grain boundary diffusion sintering. (Ref: Exner, H. E., Neck shape and limiting GBD/SD ratios in solid state sintering, *Acta Metall.*, **35**, 587–91, 1987.)

2.14. In a sintering experiment using a Cu wire wound Cu spool, Alexander and Balluffi[7] got the result shown in Figure P2.14. Explain why the variation in void area differs among the samples annealed at different temperatures.

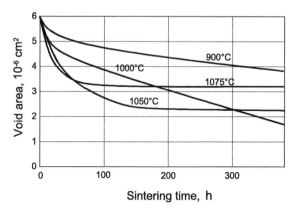

Figure P2.14

2.15. From a sintering experiment of Cu spheres on Cu plates, Kuczynski[1] deduced Cu diffusivity data, as shown in Figure P2.15. What would be the cause of the deviation of some data points from a straight line in the figure? Note that the deviation starts to occur at different temperatures for different sizes of spheres.

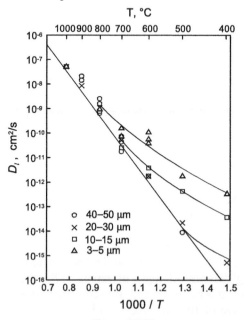

Figure P2.15

2.16. Two 10 wt%-Indium-containing Cu alloy spheres are sintered at 800°C. Explain possible microstructural changes during sintering. The solubility limit of In in Cu is ~14 wt%, and melting points of Cu and In are 1083 and 155°C, respectively.

2.17. If the capillary pressure at the neck region between two particles can cause a plastic deformation of the region, the capillary pressure must be larger than the yield strength. What condition (an equation) should be provided for the generation of dislocations by the capillary pressure? Discuss the possibility of dislocation generation by capillary pressure in real systems.

2.18. Explain the possible change in sintering kinetics of an openly porous powder compact with changing hydrostatic pressure P of an inert gas. Assume that the sintering occurs by lattice diffusion.

2.19. Consider a powder compact in which the neck growth at the initial stage sintering occurs by lattice as well as surface diffusion. Discuss the relative importance of the two mechanisms according to particle size and sintering time.

2.20. In sintering of glass spheres with a radius of 15 μm, it took 200 min at 627°C and 10 min at 677°C to get a shrinkage of 5%. Calculate the viscosity of the glass and the activation energy of sintering. The surface energy of the glass is 0.3 J/m^2.

2.21. A powder compact with a size of 1 μm is known to be sintered in one hour at 1600°C. Assuming that densification occurs by lattice diffusion with an activation energy of 500 kJ/mol, plot the dependence of the required sintering temperature against powder size for one hour sintering. Assume negligible grain growth.

2.22. The sintering rate is inversely proportional to the time required to get a constant change. An oxide MO is known to sinter by lattice as well as grain boundary diffusion at their rates, $(Rate)_l \propto (D_{l}\gamma_s V_m)/(RTG^3)$ and $(Rate)_b \propto (D_b\delta_b\gamma_s V_m)/(RTG^4)$, respectively. Given that $D_l^M > D_l^O$ and $(D_b\delta_b)^O > (D_b\delta_b)^M$, determine the dominant sintering mechanism of the oxide powders with changing particle size from very fine to very coarse.

2.23. Using appropriate data in Reference 27, construct sintering diagrams of Cu under the following ranges of experimental conditions.
   (a) Sintering diagram of $\log a$ versus $1/T$, showing the regions of dominant mechanisms for a shrinkage of 5% in the ranges of 10^{-6}–10^{-3} m in particle size and 600–1300K in temperature.
   (b) Sintering diagram of $\log(x/a)$ versus T/T_m showing the regions of dominant mechanisms and the contours of constant time for a particle size of 10^{-3} m.

2.24. Consider an ideally close-packed monosize powder compact. To explain the densification rate at a given moment an engineer assumed the driving force of densification to be the difference in total interfacial energy between the powder compact at that moment and a fully dense powder compact. Is this assumption acceptable under the condition of negligible grain growth? Explain.

2.25. Explain why the shrinkage equation derived from a two-particle model cannot be applicable to the prediction of the shrinkage of a real powder compact with a particle size distribution.

2.26. At the final stage of sintering, the densification of a powder compact occurs by both lattice and grain boundary diffusion. With an increased sintering time, which mechanism is more important for densification? Is it possible to get an answer from Herring's scaling law? Assume negligible grain growth.

2.27. According to Coble's final stage sintering model, the rate of pore volume change dV_p/dt by lattice diffusion is expressed as

$$\frac{dV_p}{dt} = -\frac{144 D_l V_m}{RTG^3}\left(\frac{r_1 r_2}{r_2 - r_1}\right)\frac{2\gamma_s}{r_1}$$

where r_1 is the pore radius and r_2 the effective diffusion distance. Assuming $r_2 \gg r_1$, derive an equation for dV_p/dt under an external pressure of 1000 atm.

2.28. If the grain growth during sintering of a powder compact satisfies the cubic law, i.e. $G^3 = Kt$, what will be the dependence of the densification rate, $d\rho/dt$, on sintering time in Coble's final stage sintering model?

2.29. Derive Eqs (5.9) and (5.10).

2.30. Figure P2.30 shows densification curves of an oxide during sintering in oxygen and in argon.

(a) Assuming no difference in grain size for the different sintering atmospheres, explain a possible cause for the similar density up to the period of sintering time t_1 in the two different atmospheres.

(b) At t_2, do you expect a difference in grain size between the oxygen and argon-sintered samples? Explain.

(c) What is a possible cause of essentially no change in the sintered density of the oxygen-sintered sample between t_2 and t_3?

(d) Is it possible to calculate quantitatively the dedensification of the argon-sintered sample?

(e) Explain possible methods to obtain fully dense sintered compacts.

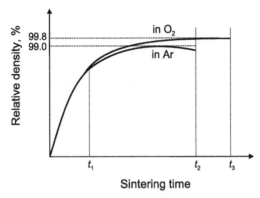

Figure P2.30

2.31. Assuming γ_s to be constant, what will be the change in densification rate with dihedral angle of a compact containing insoluble gases in isolated pores?

2.32. Kang and Yoon[40] calculated the maximum attainable density of a powder compact containing insoluble gases within isolated pores with no coalescence, as shown in Figure 5.5. Assuming that the number of pores per grain is constant, explain quantitatively how you can use this figure to predict the maximum attainable density of the same compact with a grain growth of S times.

2.33. Prove Eq. (5.12) by taking a specific system (for example, a system consisting of cubic grains and a pore at each grain boundary with a dihedral angle ϕ).

2.34. Consider a powder compact which consists of two kinds of agglomerates with different densities, ρ_h and ρ_l ($\rho_h > \rho_l$). If the relative densification rates of these agglomerates, $(d\rho/dt)/\rho$, are the same, what would happen? Discuss whether this assumption is satisfied in real sintering.

2.35. The diffusional material transport from grain boundary to neck during hot pressing is similar to the diffusional creep of polycrystalline materials.

   (a) During hot processing of a powder compact at T_1, the densification is reported to occur by Nabarro–Herring creep. If you increase the hot pressing temperature, is it possible that the powder compact is densified by Coble creep?

   (b) In the case of hot pressing, derive the equations showing the dependence of densification rate on grain size for the mechanisms of grain boundary diffusion and lattice diffusion.

   (c) Plot schematically the dependence of apparent densification rate on grain size of Al_2O_3 during hot pressing. Assume that both grain boundary diffusion and lattice diffusion are operative.

2.36. Derive Eqs (5.27) and (5.28).

2.37. Discuss the effects of particle shape on densification during pressure-assisted sintering and pressureless sintering.

2.38. Describe possible techniques that can enhance the sintering of a nitride with a high vapour pressure and explain why these techniques work.

2.39. Explain how one can measure the sintering pressure in films.

REFERENCES

1. Kuczynski, G. C., Self-diffusion in sintering of metallic particles, *Metall. Trans. AIME*, **185**, 169–78, 1949.
2. Koblenz, W. S., Dynys, J. M., Cannon, R. M. and Coble, R. L., Initial stage solid state sintering models. A critical analysis and assessment, in *Sintering Processes, Mater. Sci. Res.*, Vol. 13, Plenum Press, New York, 141–57, 1980.
3. Shewmon, P. G., *Diffusion in Solids* (2nd edition), TMS, Warrendale, PA, 84–86, 1989.
4. Frenkel, J., Viscous flow of crystalline bodies under the action of surface tension, *J. Phys. (USSR)*, **9**, 385–91, 1945.
5. Kingery, W. D. and Berg, M., Study of the initial stages of sintering solids by viscous flow, evaporation–condensation and self-diffusion, *J. Appl. Phys.*, **26**, 1205–12, 1955.
6. Balluffi, R. W. and Seigle, L. L., Effect of grain boundaries upon pore formation and dimensional changes during diffusion, *Acta Metall.*, **3**, 170–77, 1955.
7. Alexander, B. H. and Balluffi, R. W., The mechanism of sintering of copper, *Acta Metall.*, **5**, 666–77, 1957.
8. Kuczynski, G. C., Matsumura, G. and Cullity, B. D., Segregation in homogeneous alloys during sintering, *Acta Metall.*, **8**, 209–15, 1960.
9. Brett, J. and Seigle, L. L., Shrinkage of voids in copper, *Acta Metall.*, **14**, 575–82, 1966.
10. Sheehan, J. E., Lenel, F. V. and Ansell, G. S., Investigation of the early stages of sintering by transmission electron micrography, in *Sintering and Related Phenomena*, G. C. Kuczynski (ed.), Plenum Press, New York, 201–208, 1971.
11. Barrett, C. R., Nix, W. D. and Tetelman, A. S., *The Principles of Engineering Materials*, Prentice-Hall, Englewood Cliffs, New Jersey, 240–46, 1973.
12. Lenel, F. V., The role of plastic deformation in sintering, presented at the 4th Int. Symp. on Science and Technology of Sintering, Tokyo, Japan, 4–6 Nov., 1987, published in *Sintering Key Papers*, S. Sōmiya and Y. Moriyoshi (eds), Elsevier Applied Science, London, 543–64, 1990.
13. Schatt, W., Friedrich, E. and Wieters, K.-P., Dislocation activated sintering, *Rev. Powder Metall. Phys. Ceram.*, **3**, 1–111, 1984.
14. Herring, C., Surface tension as a motivation for sintering, in *The Physics of Powder Metallurgy*, W. E. Kingston (ed.), McGraw-Hill, New York, 143–79, 1951.
15. Berrin, L. and Johnson, D. L., Precise diffusion sintering models for initial shrinkage and neck growth, in *Sintering and Related Phenomena*, G. C. Kuczynski, N. A. Hooton and C. F. Gibbon (eds), Gordon and Breach, New York, 369–92, 1967.

16. Johnson, D. L., Impurity effects in the initial stage sintering of oxides, in *Sintering and Related Phenomena*, G. C. Kuczynski, N. A. Hooton and C. F. Gibbon (eds), Gordon and Breach, New York, 393–400, 1967.

17. Exner, H. E. and Bross, P., Material transport rate and stress distribution during grain boundary diffusion driven by surface tension, *Acta Metall.*, **27**, 1007–12, 1979.

18. Murr, L. E., *Interfacial Phenomena in Metals and Alloys*, Addison-Wesley, London, 187–226, 1975.

19. Humphreys, F. J. and Hatherly, M., *Recrystallization and Related Annealing Phenomena*, Pergamon, Oxford, 57–83, 1996.

20. Howe, J. M., *Interfaces in Materials*, John Wiley & Sons, New York, 297–306, 1997.

21. Sutton, A. P. and Balluffi, R. W., *Interfaces in Crystalline Materials*, Clarendon Press, Oxford, 598–654, 1995.

22. Choi, S.-Y. and Kang, S.-J. L., Sintering kinetics by structural transition at grain boundaries in barium titanate, *Acta Mater.*, **52**, 2937–43, 2004.

23. Bross, P. and Exner, H. E., Computer simulation of sintering processes, *Acta Metall.*, **27**, 1013–20, 1979.

24. Nabarro, F. R. N., Deformation of crystals by the motion of sintering ions, in *Report of a Conference on the Strength of Solids*, The Physical Society, London, 75–90, 1948.

25. Herring, C., Diffusional viscosity of a polycrystalline solid, *J. Appl. Phys.*, **21**, 437–45, 1950.

26. Coble, R. L., A model for boundary diffusion controlled creep in polycrystalline materials, *J. Appl. Phys.*, **34**, 1679–82, 1963.

27. Ashby, M. F., A first report on sintering diagrams, *Acta Metall.*, **22**, 275–89, 1974.

28. Swinkels, F. B. and Ashby, M. F., A second report on sintering diagrams, *Acta Metall.*, **29**, 259–81, 1981.

29. Herring, C., Effect of change of scale on sintering phenomena, *J. Appl. Phys.*, **21**, 301–303, 1950.

30. Nichols, F. A. and Mullins, W. W., Surface- (interface-) and volume-diffusion contributions to morphological changes driven by capillarity, *Trans. AIME*, **233**, 1840–48, 1965.

31. Noh, J.-W., Kim, S.-S. and Churn, K.-S., Collapse of interconnected open pores in solid state sintering of W-Ni, *Metall. Trans. A.*, **23A**, 2141–45, 1992.

32. Coble, R. L., Sintering of crystalline solids. I. Intermediate and final state diffusion models, *J. Appl. Phys.*, **32**, 789–92, 1961.

33. Coble, R. L. and Gupta, T. K., Intermediate stage sintering, in *Sintering and Related Phenomena*, G. C. Kuczynski, N. A. Hooton and C. F. Gibbon (eds), Gordon and Breach, New York, 423–44, 1967.

34. Zhao, J. and Harmer, M. P., Sintering kinetics for a model final-stage microstructure: A study of Al₂O₃, *Phil. Mag. Lett.*, **63**, 7–14, 1991.

35. Kang, S.-J. L. and Jung, Y.-I., Sintering kinetics at final stage sintering: model calculation and map construction, *Acta Mater.*, **52**, 4373–78, 2004.

36. Brook, R. J., Fabrication principles for the production of ceramics with superior mechanical properties, *Proc. Brit. Ceram. Soc.*, **32**, 7–24, 1982.

37. Thompson, A. M. and Harmer, M. P., Influence of atmosphere on the final-stage sintering kinetics of ultra-high-purity alumina, *J. Am. Ceram. Soc.*, **76**, 2248–56, 1993.

38. Coble, R. L., Sintering alumina: effect of atmospheres, *J. Am. Ceram. Soc.*, **45**, 123–27, 1962.

39. Paek, Y.-K., Eun, K.-Y. and Kang, S.-J. L., Effect of sintering atmosphere on densification of MgO doped Al_2O_3, *J. Am. Ceram. Soc.*, **71**, C380–82, 1988.

40. Kang, S.-J. L. and Yoon, K. J., Densification of ceramics containing entrapped gases, *J. Eu. Ceram. Soc.*, **5**, 135–39, 1989.

41. Yoon, K. J. and Kang, S.-J. L., Densification of ceramics containing entrapped gases during pressure sintering, *J. Eu. Ceram. Soc.*, **6**, 201–02, 1990.

42. Raj, R., Analysis of the sintering pressure, *J. Am. Ceram. Soc.*, **70**, C210–11, 1987.

43. Kang, S.-J. L., Comment on analysis of the sintering pressure, *J. Am. Ceram. Soc.*, **76**, 1902, 1993.

44. Svoboda, J., Riedel, H. and Zipse, H., Equilibrium pore surfaces, sintering stresses and constitutive equations for the intermediate and late stage of sintering — I. Computation of equilibrium surfaces, *Acta Metall. Mater.*, **42**, 435–43, 1994.

45. Petzow, G. and Exner, H. E., Particle rearrangement in solid state sintering, *Z. Metallkd.*, **67**, 611–18, 1976.

46. Liniger, E. and Raj, R., Packing and sintering of two-dimensional structures made from bimodal particle size distributions, *J. Am. Ceram. Soc.*, **70**, 843–49, 1987.

47. Sudre, O. and Lange, F. F., The effect of inclusions on densification: III, The desintering phenomenon, *J. Am. Ceram. Soc.*, **75**, 3241–51, 1992.

48. Lange, F. F., De-sintering, a phenomenon concurrent with densification within powder compacts: a review, in *Sintering Technology*, R. M. German, G. L. Messing and R. G. Cornwall (eds), Marcel Dekker, New York, 1–12, 1996.

49. Evans, A. G., Structural reliability: A processing-dependent phenomenon, *J. Am. Ceram. Soc.*, **65**, 127–37, 1982.

50. Lange, F. F. and Metcalf, M., Processing-related fracture origins: II, Agglomerate motion and cracklike internal surfaces caused by differential sintering, *J. Am. Ceram. Soc.*, **66**, 398–406, 1983.

51. De Jonghe, L. C. and Rahaman, M. N., Sintering stress of homogeneous and heterogeneous powder compacts, *Acta Metall.*, **36**, 223–29, 1988.

52. Weiser, M. W. and De Jonghe, L. C., Inclusion size and sintering of composite powders, *J. Am. Ceram. Soc.*, **71**, C125–27, 1988.

53. Park, S. Y., unpublished micrograph, 1989.

54. Kang, S.-J. L., Greil, P., Mitomo, M. and Moon, J.-H., Elimination of large pores during gas-pressure sintering of β' sialon, *J. Am. Ceram. Soc.*, **72**, 1166–69, 1989.

55. Kwon, S.-T., Kim, D.-Y., Kang, T.-K. and Yoon, D. N., Effect of sintering temperature on the densification of Al_2O_3, *J. Am. Ceram. Soc.*, **70**, C69–70, 1987.

56. Dobedoe, R. S., West, G. D. and Lewis, M. H., Spark plasma sintering of ceramics, *Bull. Eu. Ceram. Soc.*, **1**, 19–24, 2003.

57. Coble, R. L., Diffusion models for hot pressing with surface energy and pressure effects as driving forces, *J. Appl. Phys.*, **41**, 4798–807, 1970.

58. Arzt, E., Ashby, M. F. and Eastering, K. E., Practical applications of hot-isostatic pressing diagrams: four case studies, *Metall. Trans. A*, **14A**, 211–21, 1983.

59. Helle, A. S., Eastering, K. E. and Ashby, M. F., Hot-isostatic pressing diagrams: new developments, *Acta Metall.*, **33**, 2163–74, 1985.

60. Swinkels, F. B., Wilkinson, D. S., Arzt, E. and Ashby, M. F., Mechanisms of hot-isostatic pressing, *Acta Metall.*, **31**, 1829–40, 1983.

61. Scherer, G. W., Sintering inhomogeneous glasses—application to optical wave-guides, *J. Non-Cryst. Solids*, **34**, 239–56, 1979.

62. Bordia, R. K. and Scherer, G. W., On constrained sintering — I. Constitutive model for a sintering body, *Acta Metall.*, **36**, 2393–97, 1988.

63. Bordia, R. K. and Scherer, G. W., On constrained sintering — II. Comparison of constitutive models, *Acta Metall.*, **36**, 2399–409, 1988.

64. Rahaman, M. N. and De Jonghe, L. C., Sintering of CdO under low applied stress, *J. Am. Ceram. Soc.*, **67**, C205–207, 1984.

65. Venkatachari, K. R. and Raj, R., Shear deformation and densification of powder compacts, *J. Am. Ceram. Soc.*, **69**, 499–506, 1986.

66. Scherer, G. W., Viscous sintering under a uniaxial load, *J. Am. Ceram. Soc.*, **69**, C206–207, 1986.

67. Cheng, T. and Raj, R., Measurement of the sintering pressure in ceramic films, *J. Am. Ceram. Soc.*, **71**, 276–80, 1988.

68. McMeeking, R. M. and Kuhn, L. T., A diffusional creep law for powder compacts, *Acta Metall. Mater.*, **40**, 961–69, 1992.

69. Riedel, H., Zipse, H. and Svoboda, J., Equilibrium pore surfaces, sintering stresses and constitutive equations for the intermediate and late stages of sintering — II. Diffusional densification and creep, *Acta Metall. Mater.*, **42**, 445–52, 1994.

70. Rahaman, M. N., De Jonghe, L. C. and Brook, R. J., Effect of shear stress on sintering, *J. Am. Ceram. Soc.*, **69**, 53–58, 1986.

71. Zuo, R., Aulbach, E. and Rödel, J., Viscous Poisson's coefficient determined by discontinuous hot forging, *J. Mater. Res.*, **18**, 2170–76, 2003.

72. Zuo, R., Aulbach, E. and Rödel, J., Experimental determination of sintering stresses and sintering viscosities, *Acta Mater.*, **51**, 4563–74, 2003.

73. Bordia, R. K. and Raj, R., Sintering behavior of ceramic films constrained by a rigid substrate, *J. Am. Ceram. Soc.*, **68**, 287–92, 1985.

74. Garino, T. J. and Bowen, H. K., Kinetics of constrained film sintering, *J. Am. Ceram. Soc.*, **73**, 251–57, 1990.

75. Bang, J. and Lu, G.-Q., Densification kinetics of glass films constrained on rigid substrates, *J. Mat. Res.*, **10**, 1321–26, 1995.

76. Stech, M., Reynders, P. and Rödel, J., Constrained film sintering of nanocrystalline TiO$_2$, *J. Am. Ceram. Soc.*, **83**, 1889–96, 2000.

77. Kanters, J., Eisele, U. and Rödel, J., Co-sintering simulation and experimentation: case study of nanocrystalline zirconia, *J. Am. Ceram. Soc.*, **84**, 2757–63, 2001.

PART III
GRAIN GROWTH

The average grain size of polycrystalline materials increases as the annealing time increases and the phenomenon of grain growth is important not only in sintering but also in other materials processes. Phenomenologically, grain growth is divided into two types: normal and abnormal (sometimes also called exaggerated) grain growth. Normal grain growth is characterized by a simple and invariable distribution of relative grain sizes with annealing time, while abnormal grain growth, which occurs by the formation of some exceptionally large grains in a matrix of fine grains, shows a bimodal grain size distribution. From the point of view of chemistry and microstructure, several cases may be considered in the study of grain growth: pure materials, materials with impurity segregation at the grain boundaries, materials with second-phase particles, materials in chemical inequilibrium, etc. Here, grain growth behaviour and its theoretical basis in fully dense polycrystalline materials will be considered. Grain growth during densification of powder compacts in solid state sintering will be discussed in Chapter 11. The growth of grains dispersed in a liquid or solid matrix, so-called Ostwald ripening, will be discussed in Chapter 15.

6

NORMAL GRAIN GROWTH AND SECOND-PHASE PARTICLES

6.1 NORMAL GRAIN GROWTH

Grain growth in polycrystals is best explained in terms of a chemically pure single-phase system. Even in this case, however, the kinetics of movement varies from boundary to boundary, because the grain boundary energy varies with the grain boundary orientation and the grain boundary mobility may not be constant.[1] As a result, grain growth cannot be rigorously analysed even in this simplest system and a number of theories have been proposed.[2] For a fundamental understanding of grain growth, however, the classical theory[3,4] developed on the assumption of constant grain boundary energy is useful.

Figure 6.1 shows a typical, single-phase polycrystalline microstructure. For such a microstructure, the average grain shape is hexagonal[5] but most of the grain boundaries are curved following the number of surrounding grains. The atoms on both sides of a curved grain boundary are under different pressures across the boundary. If the atoms are in local equilibrium, the pressure difference ΔP is $2\gamma_b/R_o$, where R_o is the radius of curvature of the grain boundary. Since the growth rate of an average sized grain $\overline{G}$ must be proportional to the average velocity $\overline{v}_b$ of grain boundary movement,

$$\frac{\mathrm{d}\overline{G}}{\mathrm{d}t} = \alpha\overline{v}_b = \alpha J\Omega$$

$$= \alpha\frac{D_b^{\perp}}{kT}(\nabla P)\Omega$$

$$= \alpha\frac{D_b^{\perp}}{RT}\frac{2\gamma_b}{\overline{R}_o}\frac{V_m}{\omega} \qquad (6.1)$$

Here, α, J, Ω, $D_b^{\perp}$, ω and k are, respectively, the proportionality constant, atom flux, atom volume, atom diffusion coefficient across the grain boundary, grain

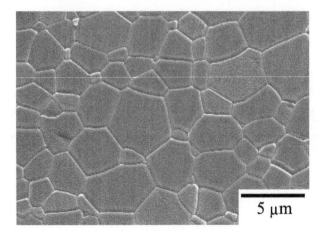

Figure 6.1. Typical microstructure of a single-phase polycrystalline material (sintered alumina).

boundary thickness, and Boltzmann constant (1.3806 J/K). For a given average grain size and grain size distribution, Eq. (6.1) can be rewritten as

$$\frac{d\overline{G}}{dt} = \frac{D_b^{\perp} 2\gamma_b V_m}{\beta RT\overline{G}\omega} \tag{6.2}$$

because the average radius of curvature is proportional to the average grain size. Here, β is a constant which includes α. Integration of Eq. (6.2) from time t_0 to t gives

$$\overline{G}_t^2 - \overline{G}_{t_0}^2 = \frac{4D_b^{\perp}\gamma_b V_m}{\beta RT\omega} t \tag{6.3}$$

Therefore, the grain growth is proportional to the square root of the annealing time. The classical model of Eq. (6.3) was derived for bulk polycrystalline materials but in the case of thin films with two-dimensional grains, the kinetics is reduced by one-half because the pressure difference across the boundary is then $\Delta P = \gamma_b/R_o$.

Classical theory explains the complicated phenomenon of grain growth in a simple manner. It is assumed that the driving force is determined only by the radius of curvature of the grain boundary and that the average grain growth rate is proportional to the average rate of grain boundary movement. The latter assumption is justified only when the grain shape and the grain size distribution are invariable during the grain growth. This condition appears to be satisfied in real microstructures that are free of abnormal grain growth.

As annealing time increases, the grain size distribution reaches a stationary state and the average grain size increases as a function of the square root of annealing time ($t^{1/2}$). However, a theoretically rigorous treatment of this phenomenon does not seem to be complete as yet.[2,6,7]

The dynamic shape change of grains during growth can be explained in terms of topology.[5] Figure 6.2 shows schematically a two-dimensional microstructure and an overlapping large hexagonal array. As the grains grow to the size of the large hexagon, the external edges (boundaries) persist while the internal edges have disappeared. In such a process, the overall change in microstructure during grain growth can be simply explained. Two types of corners are present in the microstructure in Figure 6.2; one of these is shared by two polygons (*in*) while the other has only one polygon (*out*). Denoting h_i and h_o as the numbers of each type and using the expression in Section 3.2, the total number of corners in this microstructure, C, is expressed as

$$C = \frac{\sum nP_n}{3} + \frac{1}{3}h_i + \frac{2}{3}h_o \qquad (6.4)$$

In addition, since $E_b = h_i + h_o$,

$$E = \frac{\sum nP_n}{2} + \frac{h_i + h_o}{2} \qquad (6.5)$$

Therefore, from Eq. (3.4),

$$\sum (6 - n)P_n + h_o - h_i = 6 \qquad (6.6)$$

This equation implies that after grain growth, the distribution of large grains is determined by the original grouping of small grains. It also indicates that the

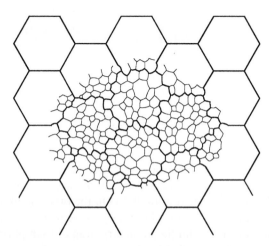

Figure 6.2. Hexagons superimposed on a group of tri-connected polygons.[5]

average shape of large grains is a hexagon as a result of the disappearance of internal small grains whose average number of edges is also 6.

Whether a specific grain is shrinking or growing during grain growth is analogous to the case of the volume changes occurring in a gas bubble of the same size by gas diffusion through the bubble wall. For simplicity, consider a two-dimensional microstructure. The rate of area change of a grain dA/dt is expressed as[8,9]

$$\frac{dA}{dt} = \frac{\pi M \gamma_b}{3}(n - 6) \tag{6.7}$$

where A is the grain area, M the grain boundary mobility and n the number of edges of the grain. This equation shows that the change of grain size is determined only by the number of edges. The equation also indicates that the area of a six-sided grain is invariable and that the rate of area change of n-sided polygons other than a hexagon is proportional to $(n - 6)$. This result is acceptable only when the grain boundary movement occurs by the diffusion of atoms across the boundary and is, therefore, continuous. In reality, however, not all grain boundary movement is continuous. A discontinuous movement, such as the instantaneous disappearance of small grains, may occur and the number of edges suddenly changes.[10] Therefore, Eq. (6.7) can provide only a limited explanation of the kinetics of grain growth. According to a recent investigation on two-dimensional microstructures,[10] the grains become unstable and disappear when the average number of edges is ~4.5. In the two-dimensional cross-section of a bulk polycrystal, however, many triangular grains are observed and these disappear continuously. Such observations suggest that the cross-sectional microstructure of a bulk polycrystal cannot be considered in terms of a simple two-dimensional array microstructure.

A number of experimental investigations on grain growth of zone-refined polycrystals have been made.[2,11] However, the reported kinetics rarely satisfy the kinetic equation (6.3) and the measured exponent is usually larger than 2. Reports of exponent values >2 in grain growth measurements occur, in general, when the grain size is large, the sample contains an appreciable amount of impurities or the annealing temperature is low. The discrepancy between the theory and experiments can be attributed to impurity drag of grain boundaries (see Chapter 7) even in zone-refined material and a threshold driving force for grain boundary movement.[11] However, the existence of a threshold driving force needs clarification. According to investigations on the effect of boundary structure, the movement of faceted boundaries needs a critical driving force.[12–15] (See Section 9.2.) On the other hand, the presence of a minimum driving force for grain boundary movement is evident in systems with second-phase particles and the boundary drag of such particles is referred to as the Zener (sometimes called Smith-Zener) effect.[16]

6.2 EFFECT OF SECOND-PHASE PARTICLES ON GRAIN GROWTH: ZENER EFFECT

When second-phase particles are present at grain boundaries as shown schematically in Figure 6.3, the particles hinder the grain growth.[16–18] The drag force of a particle against the boundary movement in Figure 6.3, F_d, is

$$
\begin{aligned}
F_d &= \gamma_b \sin\theta \times 2\pi r \cos\theta \\
&= \pi r \gamma_b \sin 2\theta
\end{aligned}
\tag{6.8}
$$

Therefore, the maximum drag force, $F_d^{\max}$, is $\pi r \gamma_b$ at $\theta = 45°$. Thermodynamically, the drag force results from a reduction in total grain boundary energy which is achieved by the boundary occupation of the second-phase particles.

If the second-phase particles are randomly distributed, the maximum drag force against grain growth can easily be calculated following Zener's proposal. Assuming that the radii of second-phase particles are constant, r, and their volume fraction is f_v, the number of particles attached to a unit area of grain boundary is

$$
2rf_v \Big/ \frac{4}{3}\pi r^3 = \frac{3f_v}{2\pi r^2}
\tag{6.9}
$$

Then, the maximum drag force against grain boundary movement per unit area of grain boundary, F_d^σ, is

$$
F_d^\sigma = \frac{3f_v \gamma_b}{2r}
\tag{6.10}
$$

and the work needed for 1 mol of atoms to move across the boundary, i.e. the grain boundary movement is

$$
W = \frac{3f_v \gamma_b V_m}{2r}
\tag{6.11}
$$

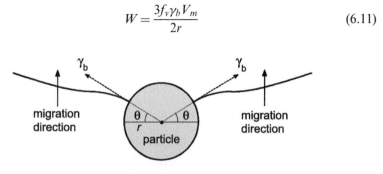

Figure 6.3. Dragging of grain boundary movement by second-phase particles: the Zener effect.

Therefore,

$$\frac{d\overline{G}}{dt} = \frac{D_b^{\perp}}{RT}\frac{1}{\omega}\left[2\gamma_b\frac{V_m}{\beta\overline{G}} - \frac{3f_v\gamma_b V_m}{2r}\right] \tag{6.12}$$

Equation (6.12) shows that the driving force for grain growth disappears when the value of the bracketed term becomes zero. In other words, there is a limiting grain size $\overline{G}_l$,

$$\overline{G}_l = \frac{4r}{3f_v\beta} \qquad \text{or} \qquad \overline{R}_l = \frac{2r}{3f_v\beta} \tag{6.13}$$

where $\overline{R}_l$ is the radius of the limiting grain size. When β is $1/2$, this equation is the original equation of Zener. According to Eq. (6.13), the limiting grain size decreases as the volume fraction of second-phase particles is increased and their size reduced. Reduction in limiting grain size, i.e. increase in grain boundary drag with reduction in particle size, is thermodynamically due to a reduction in total grain boundary area and energy with the particle size reduction.

The above calculation, however, exaggerates the grain boundary drag of second-phase particles. In reality, the particles in front of the moving grain boundary do not hinder the boundary movement but facilitate it up to their equator while those behind the boundary drag it up to and even beyond the distance of their radius from the moving boundary.[17,18] Considering this phenomenon, Louat[17] calculated the limiting grain size and found that the value of $\overline{G}_l$ is larger than that calculated from Eq. (6.13) for ratios of grain to particle radius $<10^7$.

During grain growth, second-phase particles may also grow.[19] According to Eq. (6.23), the limiting grain size increases as the second-phase particles grow. For particle growth by lattice diffusion, $r \propto t^{1/3}$ (see Section 15.2.1), and for that by grain boundary diffusion, $r \propto t^{1/4}$.[20,21] In these cases, the time dependency of $\overline{G}_l$ should be similar to that of particle growth. In reality, however, it would be difficult to experimentally confirm the dependency because the assumption of random distribution of second-phase particles may not hold during grain growth.

7

GRAIN BOUNDARY SEGREGATION AND GRAIN BOUNDARY MIGRATION

7.1 SOLUTE SEGREGATION AT GRAIN BOUNDARIES

Solute segregation at grain boundaries occurs when there are lower energy sites at the grain boundary for solute atoms than in the bulk. Consider N lattice sites with P solute atoms distributed among them and n grain boundary atom sites with p solute atoms at the grain boundary. Let E be the energy increase caused by a solute atom in the bulk and e the energy increase caused by a solute atom at the grain boundary. Then, the free energy increase due to solute atoms, G, is expressed as

$$G = pe + PE - kT[\ln n!N! - \ln(n-p)!p!(N-P)!P!] \tag{7.1}$$

Use of equilibrium conditions ($dG/dP = 0$ and $dG/dp = 0$) and Stirling's approximation ($\ln N! = N\ln N - N$ for $N \gg 1$) for Eq. (7.1) gives an equation of equilibrium distribution of solute atoms as,

$$\frac{p}{n-p} = \frac{P}{N-P}\exp\left(\frac{-(e-E)}{kT}\right) \tag{7.2}$$

Let A be the solvent atom, B the solute atom, X_A^b, X_B^b, X_A, and X_B their mole fraction at the grain boundary and in the bulk, respectively. Equation (7.2) is, then, expressed as

$$\frac{X_B^b}{X_A^b} = \frac{X_B}{X_A}\exp\left(\frac{-\Delta E}{kT}\right) \tag{7.3}$$

where $\Delta E (= e - E)$ is the free energy change due to the solute segregation at the grain boundary.

97

A number of models and theories of grain boundary segregation have been proposed.[22] The simplest of these, McLean's model,[23] is basically an application of the Langmuir-type surface adsorption to grain boundary adsorption and assumes mono-layer segregation of a single adsorbate without interference between solvent and solute atoms (no site-to-site interaction). The assumption of no interference means that for any site, the probability of solute segregation is the same, similar to the mixing assumption of the regular solution model. McLean suggested the major driving force to be the elastic strain energy around solute atoms in the lattice that results from lattice distortion.

The lattice distortion energy, $W(E_1)$, generated by a solute atom in the bulk can be calculated by using elastic continuum theory as[23,24]

$$W = \frac{24\pi K\mu r_0 r_1 (r_1 - r_2)^2}{3Kr_1 + 4\mu r_0} \tag{7.4}$$

Here, K is the bulk modulus of the solute, μ the shear modulus of the solvent, and r_0 and r_1 the radii of solvent and solute atoms, respectively. If the solute atoms are hard spheres,[24]

$$W = \frac{4\pi Y r_0^3}{1 + \nu} \left(\frac{r_1 - r_0}{r_0} \right)^2 \tag{7.5}$$

where Y is the Young's modulus and ν the Poisson's ratio.

McLean's suggestion that the elastic strain energy introduced by solute atoms is the major driving force of segregation is too simple to be justified except when the size difference between solvent and solute atoms is considerable (for example, Bi in Cu). Based on the nearest neighbour bond model, Wynblatt and his co-workers[24,25] suggested that the driving force for surface segregation in substitutional solutions should include the surface energy difference and mixing enthalpy. According to these authors, the heat of segregation of a solute atom Δh is expressed as

$$\Delta h = (\gamma_B - \gamma_A)\sigma_A + \frac{2\Delta h_m}{ZX_AX_B} \left[Z_l(X_B - X_B^b) + Z_v \left(X_B - \frac{1}{2} \right) \right]$$
$$- \frac{24\pi K\mu r_0 r_1 (r_0 - r_1)^2}{3Kr_1 + 4\mu r_0} \tag{7.6}$$

where σ_A is the atom area of the solvent, Δh_m the mixing enthalpy per atom, Z the number of nearest neighbour atoms, Z_l the number of lateral bonds of an atom, and Z_v the number of vertical bonds. The first two terms on the right-hand side of Eq. (7.6) are a part of the surface segregation energy calculated using the bond model. For grain boundary segregation, a similar idea may be adopted.[25] Equation (7.6) assumes basically no site-to-site interaction and no interdependence of the three contributions to the segregation enthalpy.

Although these assumptions would not be satisfied in real systems, they allow comparison of the relative contributions.

If the solute segregation at the interface follows the regular solution model, Eq. (7.3) is reduced to

$$\frac{X_B^b}{X_A^b} = \frac{X_B}{X_A} \exp\left(\frac{-\Delta h}{kT}\right) \tag{7.7}$$

In view of the interface energy difference, solute segregation increases as the interface energy of the solute phase decreases, following Eqs (7.6) and (7.7). Since the surface energy is, in general, 2–3 times the grain boundary energy, the effect of the interfacial energy difference should be higher for the surface than for the grain boundary. Solute segregation increases as the value of mixing enthalpy increases so that there is an increased tendency of having a miscibility gap in the phase diagram. The three contributions to segregation enthalpy in Eq. (7.6) vary with the system concerned. However, the variation of the elastic strain energy contribution is usually the largest in metals.[24,25] (For grain boundary segregation in ionic compounds, see Chapter 13.)

It has been assumed so far that segregation occurs in a monatomic layer following the McLean model. While this assumption is acceptable for dilute solutions, segregation may also occur across multi-atomic layers and, in particular, for concentrated solutions. The use of a multi-layer adsorption model (BET—Brunauer, Emmett and Teller—model) has been suggested as the most simple and useful approximation for segregation in concentrated solutions.[22]

7.2 EFFECT OF SOLUTE SEGREGATION ON GRAIN BOUNDARY MIGRATION

When the grain boundary migrates, the solute atoms segregated at the boundary are apt to remain attached to the boundary that provides them with low energy sites. In other words, the solutes have a tendency to diffuse along with the moving boundary and this solute diffusion acts as a drag force against the boundary movement. In the case of desegregation of solute atoms, a similar dragging effect results. Cahn,[26] and Lücke and Stüwe[27] have theoretically analysed the solute drag effect of a grain boundary. To calculate this drag force, the solute distribution around the moving boundary must be found.

Using mass balance and continuity conditions (Eqs (7.8) and (7.9)), Cahn[26] calculated the solute distribution as a function of distance from the grain boundary, $C(x)$, in a steady state ($v_b = $ const.)

$$\frac{\partial C}{\partial t} = -v_b \frac{\partial C}{\partial x} \tag{7.8}$$

and

$$\frac{\partial C}{\partial t} = -\frac{\partial J}{\partial x} = -\nabla J \tag{7.9}$$

When the solute concentration is low, the chemical potential of solute atoms, μ, is expressed as $\mu = kT \ln C + E(x) + E_0$. Here, E_0 is a constant and $E(x)$ is the interaction potential of the solute atom located at a distance x from the grain boundary. From the boundary conditions $dC/dx = 0$, $dE/dx = 0$ and $C(\infty) = C_\infty$ at $x = \infty$, the solute concentration at x, $C(x)$, satisfies

$$D\frac{\partial C}{\partial x} + \frac{DC}{kT}\frac{\partial E}{\partial x} + v_b(C - C_\infty) = 0 \tag{7.10}$$

because the diffusion flux J of solute atoms is $J = -(DC/kT)(d\mu/dx)$. Here, v_b is the grain boundary velocity, D the solute diffusion coefficient, E the interaction potential at a distance x from the grain boundary, C the solute concentration at x, and C_∞ the average concentration of the solute at an infinite distance from the grain boundary, i.e. in bulk grains. The solution of Eq. (7.10) shows that the solute distribution around the grain boundary is symmetric for a stationary boundary and asymmetric for a moving boundary. The solution also shows that the concentration of the solute segregated at the boundary decreases as the boundary velocity increases. A temperature increase also reduces the segregation concentration.

The drag force exerted by the segregated solutes against the boundary movement, F_b^d, is expressed as

$$F_b^d = -\int_{-\infty}^{\infty} n(x)\frac{dE}{dx}dx$$
$$= -N_v \int_{-\infty}^{\infty} [C(x) - C(\infty)]\frac{dE}{dx}dx \tag{7.11}$$

where n and N_v are the numbers of solute and solvent atoms per unit volume, respectively. Provided that $C(x)$ satisfies Eq. (7.10), an approximate solution of Eq. (7.11) gives

$$F_b^d = \frac{\alpha C_\infty v_b}{1 + \beta^2 v_b^2} \tag{7.12}$$

where α is the drag force per unit concentration of solute and per unit velocity of moving boundary when β and v_b are small enough to allow the segregated solute atoms to move along with the boundary, and β is the time required for solute atoms to diffuse one unit distance, i.e. the inverse of the drift velocity. Figure 7.1 depicts the boundary drag force with boundary velocity for high and

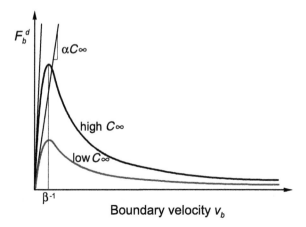

Figure 7.1. Variation of boundary drag force with boundary velocity for high and low impurity contents.[26]

low solute concentrations.[26] The slope of the curves at $v_b = 0$ is αC_∞ and F_b^d is a maximum at $v_b = \beta^{-1}$.

The total driving force F_b^t for grain boundary movement is the sum of the boundary drag force F_b^d and the force F_b^v for the movement of a grain boundary without solute segregation. Then,

$$F_b^t = F_b^v + F_b^d = \frac{v_b}{M_b^o} + \frac{\alpha C_\infty v_b}{1 + \beta^2 v_b^2}$$

$$= v_b \left(\frac{1}{M_b^o} + \frac{\alpha C_\infty}{1 + \beta^2 v_b^2} \right) \qquad (7.13)$$

where M_b^o is the boundary mobility without segregation. Figure 7.2 shows the relationship between F_b and boundary velocity.[26] As the driving force for boundary movement initially increases, the velocity of a boundary with solute segregation is much slower than that of a pure boundary because of the impurity drag. However, when the driving force is larger than a critical value, the boundary velocity increases discontinuously and approaches that of a pure material. The discontinuous increase in boundary velocity results when the driving force is too high for all the segregated solutes to move along with the boundary. On the other hand, as the driving force decreases, solute atoms are segregated again at the moving boundary, the opposite phenomenon to that observed with driving force increase. The driving force of boundary movement for solute resegregation (critical driving force for pinning) is different to that for the break-away from the solute segregation (critical driving force for break-away).

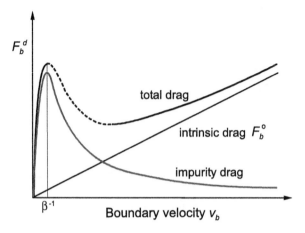

Figure 7.2. Driving force for grain boundary migration versus grain boundary velocity.[26]

For the low velocity range in Figure 7.2 ($v_b \ll \beta^{-1}$) as a result of low F_b^t and/or high solute segregation, the boundary velocity is expressed as

$$v_b = \frac{F_b^t}{(1/M_b^o) + \alpha C_\infty} \approx \frac{1}{\alpha C_\infty} F_b^t \tag{7.14}$$

This equation indicates that v_b is inversely proportional to C_∞. If the degree of segregation is the same for slow and fast diffusing solute elements, the drag effect is higher for the slow diffusing solute. On the other hand, for the high velocity range ($v_b \gg \beta^{-1}$) as a result of high F_b^t and/or low solute segregation, the boundary velocity is expressed as

$$v_b \approx M_b^o F_b^t \tag{7.15}$$

In this case relatively fast diffusing species of solute elements exhibit a higher drag effect. As the concentration of solutes decreases, the drag effect decreases as shown in Figure 7.1 and the discontinuity in boundary velocity with driving force (Figure 7.2) disappears below a critical value of solute concentration.

The reduction in boundary mobility by solute atoms (or ions) depends not only on their amount but also on their mobility. As the mobility of solute atoms decreases, the value of α in Eq. (7.14) increases. The major parameters affecting solute mobility are size and charge misfit. According to limited experimental results,[28–30] charge misfit appears to be the more important. In the case of LiF with Na, Mg and Al doping, the α values in Eq. (7.14) were measured to be $\alpha_{Al}/(7 \times 10^4) \geq \alpha_{Mg} \geq 5\alpha_{Na}$.[28] This result suggests that a small amount of an aliovalent solute, in the order of ppm, can considerably affect the grain boundary mobility. In the case of oxides (in particular those with high stoichiometry), however, this effect is less significant than in LiF.[31–34]

As explained up to the present, the effect of solute segregation on grain boundary migration has usually been explained in terms of the solute drag. However, solute segregation affects other physical properties, such as grain boundary energy and anisotropy. In particular, grain boundary energy anisotropy determines the structure of grain boundaries between faceted and rough. According to previous and recent investigations,[12–15,35] the mobility of faceted boundaries varies with the driving force, unlike a constant mobility for rough boundaries. (See Section 9.2 and Section 15.4.) Therefore, the effect of a structural change of grain boundaries by addition of a dopant can be greater than that of the boundary drag by the segregated dopant. It seems that for many cases, the two different contributions to the boundary mobility appear together. In this regard, it could be worthwhile to re-examine previous investigations on grain boundary mobility and grain growth with and without solute segregation, specifically considering the boundary structure of the materials.

8

INTERFACE MIGRATION UNDER CHEMICAL INEQUILIBRIUM

8.1 GENERAL PHENOMENA

When a polycrystalline solid becomes chemically unstable at high temperatures, equilibration reactions occur in a reasonable time span to form a new solid solution in chemical equilibrium. Sometimes the equilibrating reactions do not occur by conventional lattice diffusion but with boundary migration accompanied by boundary diffusion, forming a new solid solution behind the migrating boundaries.[36–40] This phenomenon is referred to as 'Diffusion-Induced Grain-boundary Migration (DIGM)' or 'Chemically Induced Grain-boundary Migration (CIGM)'. When a liquid film migrates in a solid–liquid two-phase system with a liquid film between grains, the phenomenon is called 'Diffusion-Induced (or Chemically Induced) Liquid Film Migration' or, simply, 'Liquid Film Migration (LFM)'. For both grain boundary and liquid film migration the term 'Diffusion-Induced (or Chemically Induced) Interface Migration (DIIM or CIIM)' is used. DIIM has been observed in a number of systems,[41–54] but is most common in systems with substitutional solute elements. Recently, however, a change in defect concentration has also been found to induce interface migration.[52,53] Figure 8.1 is a typical microstructure of DIGM showing the position of grain boundaries before and after the migration. The original position of the migrating boundaries is revealed by a thermal etching before the migration.

DIIM is observed mostly when solute elements diffuse into or from the polycrystal and where the grain boundary or liquid film provides a rapid path for material transport. The interface migrates when the solutes go into or out of the bulk grain by lattice diffusion. In general, the interface area increases with interface migration. The interface migration occurs therefore when the driving force for DIIM is larger than the energy increase caused by the reduction in the radius of curvature of the migrating interface. Unlike the

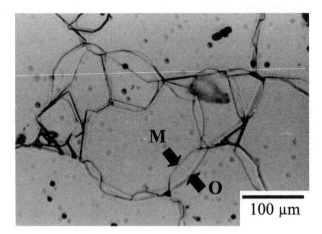

100 μm

Figure 8.1. Diffusion-induced grain boundary migration in Al_2O_3. A sintered Al_2O_3 poly-crystal annealed at 1600°C for 2 h in Fe_2O_3 vapour-containing air. Arrows with 'O' and 'M', respectively, indicate positions of grain boundaries before and after annealing.

conventionally considered solid solution formation by lattice diffusion and the consequent symmetric distribution of solutes across the interface, DIIM involves the formation of a new solid solution with an asymmetric distribution of solute atoms by interface migration and lattice diffusion at the interface.

Figure 8.2 schematically shows the solute concentration profile across the moving boundary during DIIM caused by solute addition. The chemical potential of the solute in the migrated region is essentially the same as that at the boundary. The solute concentration in front of the moving boundary in the receding grain decreases drastically from the surface of the grain. A solution of macroscopic diffusion equations for a boundary migrating at velocity v_b shows that a solute diffusion zone of thickness of about D_l/v_b exists in front of the migrating boundary.[55] The thin diffusion layer formed at the surface of the grain in shrinkage is coherent with the bulk. On the other hand, when the thickness of the migration layer is greater than a critical value, the migration layer becomes incoherent with the growing grain forming misfit dislocations, as shown in an example in Figure 8.3.[56] The transition from coherency to incoherency of the migration layer is similar to that of the precipitates in a precipitation process.

8.2 DRIVING FORCE OF DIFFUSION-INDUCED INTERFACE MIGRATION (DIIM)

After the observation of DIGM and LFM in the 1970s,[41–43] several models and mechanisms were proposed for the driving force of the phenomena.[36–38] Among them, the coherency strain model of Hillert[57] is now widely supported by some critical experiments of Yoon and others.[45,46,50–53]

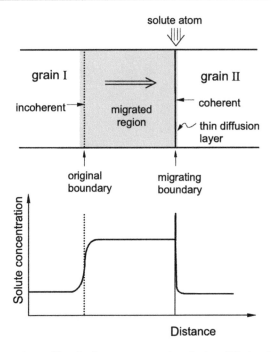

Figure 8.2. Schematic profile of solute concentration during diffusion-induced interface migration.

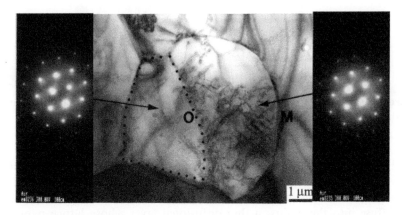

Figure 8.3. TEM micrograph showing misfit dislocations formed in a DIGM region in a $99Al_2O_3$-$1Fe_2O_3$(wt%) sample sintered at $1600°C$ in $95N_2$-$5H_2$ and then annealed at $1500°C$ in air. 'O' and 'M', respectively, indicate positions of grain boundaries before and after DIGM.[56] The diffraction patterns are from the original grain (left) and from the migrated region (right) with a [0001] axis.

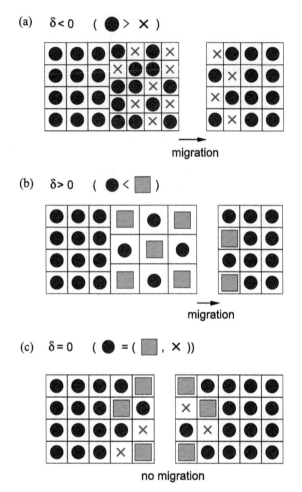

Figure 8.4. Coherency strain induced by diffusion of smaller ($\times$) or larger ($\square$) solute atoms than the host atoms ($\bullet$): (a) $\delta < 0$, (b) $\delta > 0$ and (c) $\delta = 0$.

Figure 8.4 shows schematically the coherency strain induced in a diffusional layer of solute atoms which are either smaller or larger than the solvent atoms. When the solute atoms move along the grain boundary (or a liquid film) and diffuse into the grains, a thin diffusional layer whose lattice parameter is different to that of the parent grain forms at the surface of the grain. If the layer is thin enough, it will be coherent with the parent grain, and a coherency strain energy is stored in the layer. The coherency strain energy E_c is expressed as

$$E_c = Y\varepsilon^2 \tag{8.1}$$

where Y is the coefficient of coherency strain energy and ε the coherency strain.[58,59] For cubic systems, ε is isotropic irrespective of the solute element because solute atoms are dilational centres; however, it is, in general, aniso-tropic for non-cubic systems. On the other hand, Y varies with crystallographic orientation even in cubic systems because it is a combination of elastic stiffnesses. Therefore, the coherency strain energies stored in thin diffusional layers on the surfaces of two adjacent grains are initially different to each other for all crystal systems. Yoon et al.[39] suggested that the direction of DIIM is from the grain with low strain energy to that with high strain energy. Some investigations[60–63] using single- and bi-crystals support this suggestion of DIIM direction.

As shown in Figures 8.4(a) and 8.4(b), coherency strain energy is stored in the thin diffusional layer when the solute atom is either smaller (Figure 8.4(a)) or larger (Figure 8.4(b)) than the host atom. However, when two kinds of solutes whose sizes are smaller and larger, respectively, are added at the same time, the lattice parameter of the new solid solution can become equal to that of the parent phase (lattice matching) at a specific addition ratio of the two solute species (Figure 8.4(c)). In this case, no strain energy is stored in the diffusion layer and therefore no DIIM is expected to occur. The formation of a new solid solution occurs only by lattice diffusion and its rate is a few orders of magnitude smaller than that found for DIIM. The idea of lattice matching was experimentally confirmed in the Mo-Ni-Co-Sn system, as shown in Figure 8.5.[45] This experimental result demonstrates, in turn, that the driving

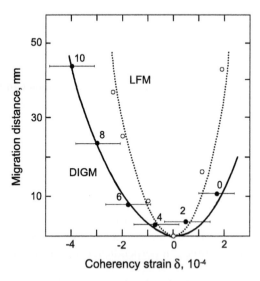

Figure 8.5. Observed variation of the average migration distance for DIGM and LFM with estimated δ in Mo-Ni specimens heat treated at 1460°C for 2 h after embedding in Mo-Ni-Co-Sn liquids of varying Co/Sn ratio.[45]

force for DIIM is not the chemical free energy of mixing but the coherency strain energy stored in a thin diffusional layer. Such an experimental demonstration has also been made in some ceramic systems.[50,51,53]

The strain energy in a diffusional layer can also vary with external pressure in the case of DIGM.[64] When an external pressure is applied to a sample, the boundary parallel to the pressure is in compression while that perpendicular to it is in tension. Therefore, if a pressure is applied to a sample where the DIGM is the result of compressive coherency strain energy, the compressive strain energy in the coherent layer parallel to the applied pressure is higher than that present without applied pressure while the compressive strain energy in the layer perpendicular to the applied pressure is lower. The result of these strain energy changes is a faster and slower migration of the boundaries parallel and perpendicular, respectively, to the compression direction.[64] Such a dependency of DIGM on external pressure also supports the view that the coherency strain energy is the driving force for DIIM.

8.3 QUANTITATIVE ANALYSIS OF DIIM

When the lattice parameters of two different phases, either bulk or thin layer on a bulk material, are different and their interface coherent, an elastic deformation of the two phases is inevitable and therefore coherency strain energy is produced. Such a coherency problem at the interface is a factor in many metallurgical and ceramic processes, such as precipitation, spinodal decomposition, thin film growth, heat treatment and interface migration. The coherency strain energy varies with crystallographic orientation, following Eq. (8.1). Cahn[58] and Hilliard[59] independently derived equations of the coherency strain energy in isotropic and cubic systems. However, a general equation of coherency strain energy applicable to any crystal system can be obtained through the transformation strain problem treated by Eshelby.[65] Hay[66] and Lee and Kang[67] have also derived general equations which can be applicable to all crystal systems. All three equations give the same result and any one of them can be used to calculate the coherency strain energy.

In the case of DIIM the coherency strain energy stored in a thin diffusional coherent layer is expressed as[67]

$$E_c = \frac{1}{2}\left[C_{ijkl}\varepsilon_{kl}\varepsilon_{ij} - (\sigma_{1'1'}\varepsilon_{1'1'} + 2\sigma_{1'3'}\varepsilon_{1'3'} + 2\sigma_{1'2'}\varepsilon_{1'2'})\right] \qquad (8.2)$$

Here C_{ijkl} is the elastic stiffness of the diffusion layer, ε_{ij} the elastic strain due to the formation of a new solid solution, and $\sigma_{i'j'}$ and $\varepsilon_{i'j'}$, respectively, the stress and strain relaxed in the direction perpendicular to the surface of the diffusion layer.

Figure 8.6[62] shows the calculated coherency strain energy map (CSEM) of rhombohedral Al_2O_3 with Fe_2O_3 as a solute from Eq. (8.2). In this figure the maximum coherency strain energy appears on the $(0001)C$ plane in terms of the

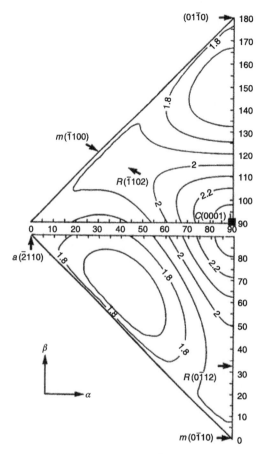

Figure 8.6. Coherent strain energy map (in MJ/m³) of Al₂O₃/Al₂O₃(Fe₂O₃) for an Fe₂O₃ concentration of 5 mol% (with $\varepsilon_c/\varepsilon_a$ of 0.94) in a coherent diffusion zone.[62] The axes represent the interplanar angles between the surface normal of the considered plane and those of an $a(\bar{2}110)$ plane (α) and an $m(0\bar{1}10)$ plane (β).

hexagonal coordination system and the minimum on the $(01\bar{1}2)q$ plane. A series of experimental investigations[61–63] using various single crystals and polycrystals have shown that the direction of DIIM was from the crystal with low E_c to that with high E_c, consistent with predictions based on the calculated CSEMs. When the difference in E_c between two crystals was not appreciable, a zigzag migration of the boundary was observed. This result may imply that in the case of inappreciable difference in E_c, local arrangement and diffusion of atoms can affect the migration direction. The change in curvature with local migration may also play a role in determining the migrating boundary shape.

The measured and estimated coherency strains in DIGM were in the range between 10^{-3} and 10^{-5} for metals as well as ceramics. However, since the stiffness of ceramics is, in general, higher than that of metals, ceramics usually have higher coherency strain energy than metals. The variation in boundary velocity with solute concentration can be predicted by estimating the coherency strain energy provided that the boundary mobility is known. However, the boundary mobility is usually not a constant but varies with boundary orientation and type (faceted or rough).[1,12–15,68,69] (See Section 9.2.1.) A recent investigation[70] showed that DIGM was suppressed in $BaTiO_3$ by a structural transition from rough to faceted. This result can be attributed to the considerable reduction of boundary mobility by the structural transition. Comparing LFM and DIGM, the velocity of LFM is higher than that of DIGM under the same driving force.[45,52,71] This result may suggest that the mobility of a liquid film is higher than that of a grain boundary. The measured boundary velocity is dependent on the system, coherency strain, temperature, etc., but it is usually in the range between $1\,\mu$m/h and $10\,\mu$m/h.

When the thickness of the front diffusion zone given by D_l/v_b increases to such a large value that coherency cannot be maintained, the diffusion layer becomes incoherent with the bulk (coherency breaking). For a cubic lattice the coherency strain at breaking $|\varepsilon_o|$ is expressed as[47]

$$|\varepsilon_o| = \frac{bv_{bc}}{4\pi(1+v)D_l}\left(\ln\frac{D_l}{bv_{bc}}+1\right) \tag{8.3}$$

where v is Poisson's ratio, b Burgers vector, and v_{bc} the critical boundary velocity for coherency breaking. When coherency breaking occurs, the boundary does not migrate any more and new solid solution forms only by lattice diffusion from the grain boundary into the bulk grains. This process is a few orders of magnitude slower than solution formation by boundary migration. As can be seen in Eq. (8.3), the occurrence or otherwise of coherency breaking is determined not only by lattice diffusivity but also by migration velocity which is in turn determined by coherency strain energy and boundary mobility.

8.4 MICROSTRUCTURAL CHARACTERISTICS OF DIIM AND ITS APPLICATION

In polycrystals, DIGM occurs either unidirectionally or in a corrugated manner and results in a variation in microstructure, as shown in Figures 8.1 and 8.7. During migration faceting of some boundaries sometimes appears (Figures 8.1 and 8.7(a)); the boundary faceting is related to a marked anisotropy of the boundary velocity in the shrinking grain.[49] (The observed parallelism in faceted boundaries (A to a, B to b(b′), C to c(c′)) of a shrinking grain G in Figure 8.7(a) can be another experimental support for the coherency strain energy being the dominant driving force for DIGM.[49]) Migration reversal is also observed, which is attributed to the coherency breaking of the

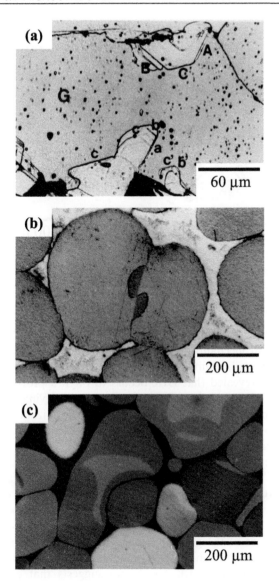

Figure 8.7. Various microstructures observed after DIIM: (a) faceted grain boundaries (Al_2O_3-Fe_2O_3),[49] (b) zigzag migration in Mo-Ni,[72] and (c) migration reversal of liquid films in Mo-Ni.[73]

front diffusion layer. In this case the boundary is at its original position but the regions of the new solid solution already formed remain within the grains, as in Figure 8.7(c).

When the difference in concentration between the parent phase and a new solid solution is high, diffusion-induced recrystallization (DIR—also called

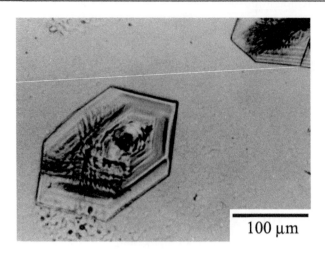

100 μm

Figure 8.8. Diffusion-induced recrystallization in an Al_2O_3 single crystal by Fe_2O_3.[80] The hexagonal grain is a recrystallized grain on an (0001) Al_2O_3 single crystal. Another recrystallized grain is also seen at the upper right corner of the figure.

chemically induced recrystallization (CIR)) can occur,[48,49,74-80] as shown in Figure 8.8. Several mechanisms have been proposed for the recrystallization under chemical inequilibrium.[74-77] In some recent investigations on TiC[78] and Al_2O_3,[79,80] it was revealed that DIR occurs through a process similar to the recrystallization of plastically deformed materials involving the formation of many dislocations within a new solid solution, polygonization and formation of new grain boundaries. The growth of recrystallized grains, however, was observed to occur by DIGM.[80]

DIGM sometimes can promote grain growth or abnormal grain growth in polycrystals.[54,70,81] Recently, Lee *et al.*[54] showed that grain growth was enhanced by DIGM under a local inhomogeneity in chemical composition. In this case the driving force for grain growth was suggested to be the sum of the coherency strain energy for DIGM and the capillary energy due to the grain boundary curvature of the grains. Figure 8.9 shows the estimated coherency strain energy and capillary energy as a function of the radius of grain boundary curvature in $BaTiO_3$ ($PbTiO_3$). The figure suggests that coherency energy dominates unless the radius of grain boundary curvature is smaller than a few tenths of a micron. Since the curvature radius is larger than the grain radius, the grain size must be smaller than the curvature value for the capillary energy to become comparable to the coherency strain energy.

During the sintering of mixed powder compacts, alloying and sintering proceed simultaneously. Since the usual sintering temperatures of materials are much lower than their melting temperatures, alloying of elemental powders can be achieved by DIGM rather than by lattice diffusion. The alloying is then much faster than expected even for coarse powders. When a liquid phase is

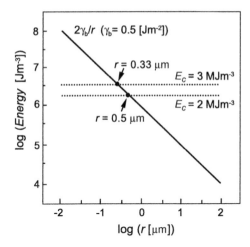

Figure 8.9. Comparison between capillary energy ($2\gamma_b/r$, r: the average radius of grain boundary curvature) and calculated coherency strain energies stored in a $(Ba_{0.8}Pb_{0.2})TiO_3$ layer on $BaTiO_3$. Specific grain boundary energy is assumed to be $0.5\,J/m^2$.[54]

present, sometimes the alloying results in fast formation of a core/shell structure.[82] The results of Lee et al.[54] suggest that the growth of solid particles can occur extensively by DIGM at the early stage of sintering, which suppresses densification. It also suggests that any chemical inhomogeneity has to be minimized in order to prevent extensive grain growth and hence improve sinterability of mixed powder compacts.

The effect of DIIM on physical properties is expected to be considerable because DIIM can drastically change the microstructure. Some recent investigations show considerable effects of DIIM on mechanical and dielectric properties in Al_2O_3-,[56] $Pb(Mg_{1/3}Nb_{2/3})O_3$-,[83] and $SrTiO_3$-based[84] materials. In the case of $Al_2O_3(Fe_2O_3)$,[56] DIGM at the surface region which resulted in the formation of many misfit dislocations in the migration layer and the corrugation of grain boundaries changed the fracture mode of the surface region from intergranular to transgranular and considerably improved the cyclic properties. Unlike the case of alumina, suppression of DIGM and LFM remarkably improved the dielectric properties of $SrTiO_3$-based boundary layer capacitors,[52,84,85] which consist of oxidized layers between conductive $SrTiO_3$ grains. The improvement of dielectric properties was due to the minimization of the thickness of the oxidized dielectric layer through suppression of the migration according to the scheme shown in Figure 8.4(c). These two examples of migration enhancement and suppression for improving mechanical and dielectric properties may be typical ones. At this stage, however, investigations on the effect of DIIM on physical properties are very limited and it would be useful to study the effect of DIIM in other systems.

9

ABNORMAL GRAIN GROWTH

Abnormal grain growth (sometimes also called exaggerated grain growth) is a coarsening type of microstructure where some (or a few) large grains grow unusually quickly in a matrix of fine grains with a very slow growth rate. In terms of microstructure, grain size distribution is bimodal, unlike in normal grain growth with a unimodal distribution. Sometimes, however, a distinction between normal and abnormal grain growth is ambiguous. It is also difficult to define a kinetic condition for abnormal grain growth. A criterion for the occurrence of abnormal grain growth was proposed to be

$$\frac{\mathrm{d}}{\mathrm{d}t}\left(\frac{G}{\overline{G}}\right) > 0 \qquad (9.1)$$

where G is the grain size of a specific grain and $\overline{G}$ the average grain size.[86] However, this condition is not sufficient to define abnormal grain growth because there is a limit to grain growth in reality due to, for example, impingement of large abnormal grains with each other. In addition, the meaning of an average size is ambiguous because only some large grains grow quickly while the other fine grains remain almost unchanged in size. However, when a normalized size distribution is not changing with annealing time, the growth can be defined as normal. On the other hand, a broadening of a normalized distribution, and in particular formation of a bimodal distribution, with annealing time can be considered to be a characteristic of abnormal grain growth.

Figure 9.1 shows an example of the microstructure of abnormal grain growth. There are several grains whose size is a few tens and hundreds of times that of the fine matrix grains. Such an abnormal grain growth is observed not only in single-phase systems but also in multi-phase systems. For a single-phase system without a liquid, the classical explanation of abnormal grain growth concerns non-uniform distribution of second-phase particles or solutes. Under the non-uniform distribution, normal grain growth may locally be suppressed

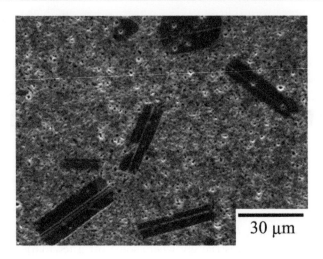

Figure 9.1. Typical microstructure of abnormal grain growth in 0.1 mol% TiO_2-excess $BaTiO_3$ sintered at 1250°C for 24 h in air.

and abnormal grains may form. The presence of a few large grains in a matrix of fine grains was also suggested to be a cause of abnormal grain growth.

In reality, however, a number of different examples are available and these may be categorized into three types:

(i)   materials containing second-phase precipitates or impurities of high concentrations,

(ii)   materials with a high anisotropy in interfacial energy, for example, solid/liquid interfacial energy or grain boundary energy in the bulk and surface energy in thin films, and

(iii)   materials in high chemical inequilibrium.

For any case, abnormal grain growth is a result of very high local rates of interface migration. A basic understanding of abnormal grain growth is just at its beginning. A number of computer simulations[86–91] have been made but theoretical analyses appear incomplete as yet.

9.1 PHENOMENOLOGICAL THEORY OF ABNORMAL GRAIN GROWTH IN SINGLE-PHASE SYSTEMS

Phenomenologically, abnormal grain growth is characterized by fast growth of a few large grains in a matrix of fine grains of which the size change with annealing time is almost nil or at least negligible compared with that of the abnormal grains. Under this condition, the growth rate of the abnormal grains,

dG_a/dt, is expressed as[4]

$$\frac{dG_a}{dt} = \frac{D_b^{\perp}}{RT} \frac{2\gamma_b}{\beta\overline{G}_m} \frac{V_m}{\omega} \tag{9.2}$$

where $\overline{G}_m$ is the average size of the matrix grains. Upon integration,

$$\overline{G}_{a,t} - \overline{G}_{a,t_o} = \frac{2D_b^{\perp}\gamma_b V_m}{\beta RT\overline{G}_m\omega} t \tag{9.3}$$

showing that the average size of abnormal grains linearly increases with annealing time. In some materials with non-uniform distribution of second-phase particles or impurities, abnormal grain growth can occur.[11] However, in other materials, abnormal grain growth did not occur under similar conditions. On the other hand, abnormal grain growth was observed in some materials with a high purity.[92,93] These results would suggest that the non-uniform distribution of second-phase particles or impurities is not the direct cause of abnormal grain growth.

9.2 INTERFACIAL ENERGY ANISOTROPY AND ABNORMAL GRAIN GROWTH

9.2.1 Single-Phase Systems

The possibility of abnormal grain growth has been predicted by computer simulations when anisotropy in grain boundary energy or grain boundary mobility is high.[87,88] Surface energy anisotropy was also predicted to be a cause of abnormal grain growth in thin films with a two-dimensional microstructure.[89] These simulations, however, were made under certain assumptions about the energy or mobility anisotropy. The physical basis of the anisotropy incorporated in the calculation was not clear.

In recent experimental investigations,[94–96] a strong correlation has been found between grain boundary structure, faceted or rough, and grain growth mode. Figure 9.2 shows an example of the two types of boundaries, faceted and rough, in the same material (in this case TiO_2-excess $BaTiO_3$).[94] The two types were distinguished at the atomic scale as well.[96,97] When grain boundaries of materials investigated were faceted, abnormal grain growth occurred.[94–96,98–100] On the other hand, for the same materials but with rough boundaries, normal grain growth occurred. These experimental results suggest that grain boundary faceting is a necessary condition for abnormal grain growth in single-phase systems, as in the case of two-phase systems. (See Section 9.2.2. and Section 15.4.)

The observed abnormal grain growth in materials with faceted boundaries can be explained in terms of the variable mobility of a facet which would move by lateral movement of boundary steps (so-called step growth mechanism or

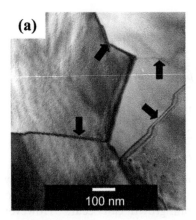 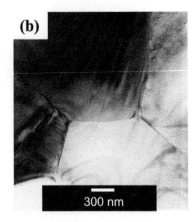

Figure 9.2. Two types of boundaries: (a) faceted and (b) rough, observed in 0.1 mol% TiO$_2$-excess BaTiO$_3$. Sintered at 1250°C for 10 h in air and then annealed (a) in air and (b) in H$_2$ at 1250°C for 48 h.[94] The arrows indicate faceted boundaries.

atom shuffle mechanism).[12,101–103] For the lateral movement of faceted boundaries, Yoon et al.[13] suggested that the mobility was not constant but varied with the driving force, as in the case of faceted solid/liquid interfaces (see Section 15.4). Under this condition, the boundaries with a driving force larger than a critical value are expected to move much faster than those with a driving force smaller than the critical value, resulting in abnormal grain growth. On the other hand, if all of the boundaries have driving forces smaller than the critical value, negligible grain growth is expected to occur. This prediction has been confirmed in the BaTiO$_3$ system.[14,15]

The mobility of faceted boundaries is also dependent on boundary defects. In the case of BaTiO$_3$, {111} twins induced and enhanced abnormal grain growth.[94,96] On the other hand, dislocations at boundaries in SrTiO$_3$ did not enhance the boundary mobility,[97] unlike the enhanced mobility of solid/liquid interfaces by dislocations.[97,104] In this case, however, the boundary was not fully faceted (~35% faceted). According to a recent investigation in BaTiO$_3$,[105] it seems that dislocations can also enhance the boundary mobility if the boundary is well faceted.

9.2.2 Two-Phase Systems

When the anisotropy in solid/liquid interfacial energy is high in solid/liquid two-phase systems, the shape of grains in a liquid matrix is polyhedral with faceted interfaces. All of the systems showing abnormal grain growth exhibited faceted grains, for example, Al$_2$O$_3$,[98,106] WC-Co,[107] Si$_3$N$_4$,[108] and SiC.[109] Figure 9.3 shows a microstructure of abnormal grains formed in impure Al$_2$O$_3$ with liquid films between grains.[110] The grain shape is mostly faceted.

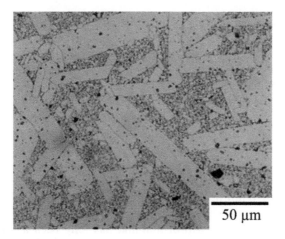

Figure 9.3. Abnormal grain growth of alumina in the presence of a liquid phase.[110] Al_2O_3 powder compact (~0.05 wt% anorthite) sintered at 1580°C for 12 h in air. (Reprinted with permission of the American Ceramic Society, www.ceramics.org.)

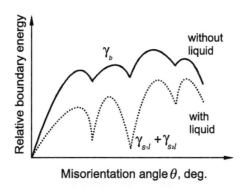

Figure 9.4. Schematic showing the variation of interfacial energy in Al_2O_3 with or without a liquid phase.[98] The interfacial energy represents $\gamma_{ss}(\gamma_b)$ or $(\gamma_{s_1l} + \gamma_{s_2l})$ of a two-dimensional Al_2O_3 crystal depending on the orientation in 360° (similar to the γ-plot of a crystal).

However, without liquid (for a highly pure Al_2O_3), the grain shape was mostly isotropic and abnormal grain growth did not occur.[32,98] These results suggest that the anisotropy of grain boundary energy is lower than that of the solid/liquid interface in Al_2O_3, as schematically shown in Figure 9.4.[98] The wetting of grain boundaries by a small amount of liquid further suggests that the grain boundary energy is higher than the two solid/liquid interfacial energies which are not constant as conventionally considered.

For faceted grains in a liquid matrix, the dihedral angle is not uniquely defined because of torque on a facet,[111,112] contrary to the case of rounded grains. When two faceted grains are in contact with a certain angle of crystal-lographic orientation in a liquid matrix, the shape of the contact boundary in equilibrium must be that of the minimum interfacial energy. Kim *et al.*[113] observed the grain boundary with a liquid film formed between two (0001) single crystals of alumina with a misorientation of ~3.5°. Figure 9.5 shows the observed boundaries with (a) and without (b) a liquid film. The boundary with a liquid film consists of faceted planes with steps while the boundary without it contains regularly spaced dislocations. This result shows that when two faceted grains come into contact and form a boundary in a liquid matrix, the boundary can consist of two types of boundaries with and without a liquid film. The formation of such a complex boundary was explained using two overlapped sets of Wulff nets with the same crystallographic misorientation of the crystals.[113]

The abnormal grain growth of alumina has long been a subject of grain growth research. According to recent investigations,[98,106,110,114,115] when a liquid phase, even a small amount, forms due to the presence of impurities, grains are faceted and abnormal grain growth can occur. Hong *et al.*[98]

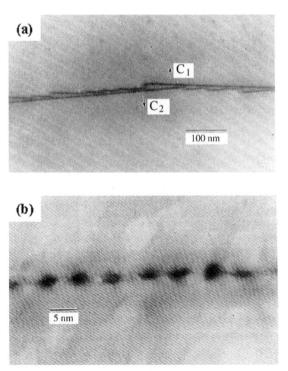

Figure 9.5. TEM micrographs showing Al_2O_3 grain boundaries (a) wetted and (b) non-wetted with a glassy phase.[113] The wetted boundaries are faceted. (Reprinted with permission of the American Ceramic Society, www.ceramics.org.)

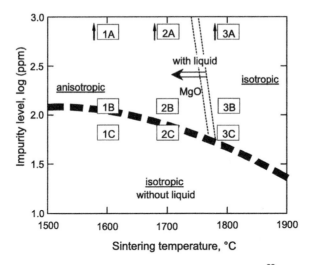

Figure 9.6. Schematic map for grain morphologies in sintered Al_2O_3.[98] The numbers in the figure, such as IA, correspond to the samples with different amounts of impurities.

prepared three kinds of Al_2O_3 compacts with different concentrations of impurities (~0.5, >0.01 and <0.01 wt%) and sintered at 1600 (10 h), 1700 (5 h) and 1800°C (4 h). Figure 9.6[98] shows the summary of their microstructure observation of the sintered samples. In samples without a liquid phase, the grain shape is isotropic and no abnormal grain growth occurs. However, when a liquid phase forms due to a high concentration of impurities, the grain shape is faceted and abnormal grain growth occurs at low annealing temperatures. The edges of faceted grains become rounded with an increase in annealing temperature and normal grain growth occurs because of an increased contribution of entropy (a reduction of step free energy*). Addition of MgO reduces the interfacial energy anisotropy and, therefore, reduces the experimental region for faceted interfaces.[110,114] Critical amounts of CaO, SiO_2 and MgO to induce and to suppress abnormal grain growth were also measured.[106,116] According to the results, when the amounts of CaO and SiO_2 are within their solubility limits, normal grain growth occurs, as schematically shown in Figure 9.6. On the other hand, abnormal grain growth occurs if the amount is above the solubility limit. The critical amount of MgO that can eliminate the interfacial energy anisotropy caused by CaO impurity and prevent abnormal grain growth has also been measured to be about the same as the amount of CaO.[116] Such a quantitative measurement, however, has rarely been made for other systems.

Recently, some attempts have been made to explain abnormal grain growth in a liquid matrix.[90,91,108,117] The explanation is based on theoretical and

---

*On this subject, see Section 15.4.

experimental results of the growth of a faceted single crystal from a liquid that showed the existence of a critical driving force for growth.[118–121] According to the explanation, only some large grains having driving forces larger than a critical value can undergo significant growth in a polycrystal with a grain size distribution, resulting in the formation of abnormal grains. On the other hand, if the driving force of large grains is smaller than a critical value, abnormally large grains do not form and normal grain growth occurs. (For details on this subject, see Section 15.4.)

9.3 ABNORMAL GRAIN GROWTH IN CHEMICAL INEQUILIBRIUM

Abnormal grain growth can also occur in high chemical in equilibrium.[54,81,122,123] Figure 9.7[81] is an example obtained after annealing of a sintered $YBa_2Cu_3O_x$ compact in contact with a $DyBa_2Cu_3O_x$ packing powder at 950°C for 20 h in O_2. In this microstructure, millimetre size abnormal grains are visible. Another example can be found in the Y-SiAlON system. When Y-β-SiAlON grains form from Y-α-SiAlON in an oxynitride glass which is in equilibrium with Y-β-SiAlON, abnormally large grains can also form.[122] Such an abnormal growth of grains in chemical inequilibrium has also been observed in Al_2O_3.[123] Although the mechanism of the observed phenomena is as yet unclear, the phenomena seem to be related to DIIM, as observed in $BaTiO_3$,[54] that can locally and considerably increase the driving force for grain growth (see Section 8.4).

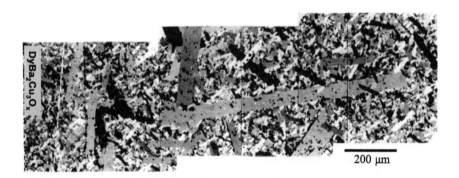

200 µm

Figure 9.7. Abnormally grown Y(Dy)Ba$_2$Cu$_3$O$_x$ superconducting grains observed after heat treating a sintered YBa$_2$Cu$_3$O$_x$ specimen packed in DyBa$_2$Cu$_3$O$_x$ at 950°C for 20 h in O$_2$.[81]

PROBLEMS

3.1. Compare and explain the normal grain growth observed in a polycrystalline pure material and a material with second-phase particles, and the classically explained abnormal grain growth.

3.2. For ionic compounds, what should be the values of $D_b^{\perp}$ and V_m in the grain growth equation (6.1)?

3.3. Prove Eq. (6.7).

3.4. Draw schematically and explain the change in drag force against grain growth caused by second-phase particles with a decreasing dihedral angle from 175°. Assume that the grain boundary energy γ_b is constant.

3.5. Consider a polycrystal with uniformly distributed second-phase particles of a fine size. What will be the variation of grain size with annealing time of the polycrystal for the following cases?
(a) No growth of the second-phase particles.
(b) Growth of the second-phase particles by lattice diffusion.
(c) Growth of the second-phase particles by grain boundary diffusion.
Draw schematic figures and explain.

3.6. Consider two sintering powder compacts of the same relative density of 96% but with and without uniformly distributed second-phase particles. Compare the densification rates of the compacts.

3.7. Calculate the degree of Ca^{2+} segregation at an MgO grain boundary assuming that the elastic strain energy is the only driving force of the segregation. Assume the Young's modulus and the Poisson's ratio of MgO to be $3 \times 10^{11} \, N/m^2$ and 0.3, respectively. The ionic radii of Mg^{2+} and Ca^{2+} are 0.72 and 1.0 Å, respectively.

3.8. Derive the solute concentration at a grain boundary, C_b,

$$C_b = \frac{C \exp(-\Delta E / RT)}{1 - C + C \exp(-\Delta E / RT)}$$

where C is the solute concentration in the bulk and ΔE the free energy change of solute segregation at the grain boundary.

3.9. Draw a schematic figure and explain the grain growth velocity as a function of grain size from 0.01 to 100 μm for a single-phase system with high solute segregation at the grain boundary. Assume normal grain growth at any grain size.

3.10. Consider a solid solution containing two kinds of solutes with very fast and very slow diffusivities perpendicular to the grain boundary. Assuming a high solute segregation at the grain boundary and the same interaction potential of the grain boundary for the two kinds of solutes, draw schematically and explain the variation of solute drag force with grain size.

3.11. Explain the variation of solute segregation at grain boundaries with increasing temperature. Draw schematically and explain the variation in grain boundary velocity with temperature for pure, slightly impure and highly impure materials.

3.12. Draw schematically and explain the variation in groove angle ϕ formed at a symmetric tilt grain boundary with tilt angle θ. As the temperature increases, how does the relationship between ϕ and θ change approximately? Assume invariable surface energy.

3.13. Explain on a figure the variation in mobility of a symmetric tilt grain boundary with tilt angle θ in an impure material. As the temperature increases, how does the grain boundary mobility change with θ?

3.14. The grain boundary mobility of a pure material is $D_b^{\perp}/kT$. Derive the grain boundary mobility for the low velocity limit of an impure material with high grain boundary segregation. Assume the grain boundary segregation follows the McLean model.

3.15. After a grain growth experiment using two kinds of fully dense polycrystals with 99.8 and 99.999% purity, an engineer found that the activation energy for grain growth was different between the two polycrystals. Which one has a higher activation energy? Answer with the species and process that affect the activation energy in both cases.

3.16. Explain how the migration direction is determined at the beginning of diffusion-induced grain boundary migration.

3.17. Diffusion-induced grain boundary migration occurs within a certain temperature range for a given material. Why?

3.18. Explain a possible process for the dissolution of W in a Ni melt.

3.19. Consider a heat treatment of metal A in contact with metal B, as shown in Figure P3.19, below their melting points. Assuming a finite solubility of B in A, draw schematic figures and explain the variation in the concentration of B in A with heat treatment temperature and time.

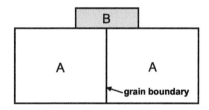

Figure P3.19

3.20. Liquid film migration occurred when solute atoms were added to a liquid phase sintered body. If the migration direction of some liquid films was reversed and the films went back to their original position during annealing, what would be the cause of this migration reversal? Knowing that the principal radii of curvature of a liquid film were r_1 and r_2 just before its migration reversal, estimate the driving force of the liquid film migration.

3.21. Consider a perfectly coherent diffusion layer on a large single crystal surface.
(a) explain the state of the elastic stress in the diffusion layer, and
(b) calculate the elastic stress and strain, and also the elastic strain energy in the layer for a cubic crystal system.

3.22. Consider a partially coherent thin β layer on an α single crystal. Given the distance between the misfit dislocations formed at the α/β interface to be d and the intrinsic (stress-free) lattice parameters of α and β to be a^{α} and a^{β}, respectively, derive the coherency strain ε in the β layer.

3.23. Show schematically and explain a diagram of Gibbs free energy versus composition for liquid film migration.

3.24. Discuss whether the grain boundary migration during discontinuous precipitation or dissolution is basically different from DIGM.

3.25. The grain boundary structure of polycrystalline $BaTiO_3$ varies with the oxygen partial pressure, p_{O_2}: faceted and rough under a high and a low p_{O_2}, respectively.[94] Assuming the same lattice and grain boundary diffusivities of solute ions in $BaTiO_3$ polycrystals with different grain boundary structures, do you expect the same degree of DIGM in two different $BaTiO_3$ samples with faceted and rough grain boundaries? Explain.

3.26. Consider fine and coarse powders of the same chemical composition. During sintering of the powders a fine powder compact showed abnormal grain growth while a coarse powder compact showed normal grain growth. Explain why the grain growth behaviour was different between the two powders.

3.27. For a system where abnormal grain growth usually occurs during sintering, what possible measures can be taken to suppress abnormal grain growth in view of the structural change of grain boundaries?

3.28. Explain in detail the formation of abnormal grains in chemical inequilibrium in terms of a DIIM process.

REFERENCES

1. See for example:
   (a) Gleiter, H. and Chalmers, B., *High-Angle Grain Boundaries*, Pergamon Press, Oxford, 127–78, 1972.
   (b) Humphreys, F. J. and Hatherly, M., *Recrystallization and Related Annealing Phenomena*, Pergamon, Oxford, 85–126, 1996.
   (c) Gottstein, G. and Shvindlerman, L. S., *Grain Boundary Migration in Metals: Thermodynamics, Kinetics, Applications*, CRC Press, Boca Raton, FL, 125–285, 1999.
2. See for example:
   (a) Atkinson, H. V., Theories of normal grain growth in pure single phase system, *Acta Metall.*, **36**, 469–91, 1988.
   (b) Humphreys, F. J. and Hatherly, M., *Recrystallization and Related Annealing Phenomena*, Pergamon, Oxford, 281–325, 1996.
3. Burke, J. E. and Turnbull, D., Recrystallization and grain growth, *Prog. Metal Phys.*, **3**, 220–91, 1952.
4. Brook, R. J., Controlled grain growth, in *Ceramic Fabrication Processes*, F. F. Y. Wang (ed.), Academic Press, New York, 331–64, 1976.
5. Smith, C. S., Some elementary principles of polycrystalline microstructure, *Metall. Reviews*, **9**, 1–48, 1964.
6. Hillert, M., On the theory of normal and abnormal grain growth, *Acta Metall.*, **13**, 227–38, 1965.
7. Pande, C. S., and Dantsker, E., On a stochastic theory of grain growth — IV, *Acta Metall. Mater.*, **42**, 2899–903, 1994.
8. von Neumann, J., in *Metal Interfaces*, Am. Soc. Metals, Cleveland, 108–10 (discussion), 1952.
9. Mullins, W. W., Two-dimensional motion of idealized grain boundaries, *J. Appl. Phys.*, **27**, 900–904, 1956.
10. Fradkov, V. E., Glicksman, M. E., Palmer, M. and Rajan, K., Topological events in two-dimensional grain growth: experiments and simulations, *Acta Metall. Mater.*, **42**, 2719–27, 1994.
11. Martin, J. W. and Doherty, R. D., *Stability of Microstructure in Metallic Systems*, Cambridge University Press, Cambridge, 221–43, 1976.
12. Gleiter, H., Theory of grain boundary migration rate, *Acta Metall.*, **17**, 853–62, 1969.
13. Yoon, D. Y., Park, C. W. and Koo, J. B., The step growth hypothesis for abnormal grain growth, in *Ceramic Interfaces 2*, H.-I. Yoo. and S.-J. L. Kang (eds), Institute of Materials, London, 3–21, 2001.

14. Jung, Y.-I., Choi, S.-Y. and Kang, S.-J. L., Grain growth behavior during stepwise sintering of barium titanate in hydrogen gas and air, *J. Am. Ceram. Soc.*, **86**, 2228–30, 2003.

15. Choi, S.-Y. and Kang, S.-J. L., Sintering kinetics by structural transition at grain boundaries in barium titanate, *Acta Mater.*, **52**, 2937–43, 2004.

16. Zener, C., Private communication to C. S. Smith, Grains, phases and interfaces: an interpretation of microstructures, *Am. Inst. Min. Metall. Engrs*, **175**, 15–51, 1949.

17. Louat, N., The resistance to normal grain growth from a dispersion of spherical particles, *Acta Metall.*, **30**, 1291–94, 1982.

18. Manohar, P. A., Ferry, M. and Chandra, T., Five decades of the Zener equation, *ISIJ Inter.*, **38**, 913–24, 1998.

19. French, J. D., Harmer, M. P., Chan, H. M. and Miller, G. A., Coarsening-resistant dual-phase interpenetrating microstructures, *J. Am. Ceram. Soc.*, **73**, 2508–10, 1990.

20. Greenwood, G. W., Particle coarsening, in *The Mechanism of Phase Transformations in Crystalline Solids*, Institute of Metals, London, 103–10, 1969.

21. Kirchner, H. O. K., Coarsening of grain-boundary precipitates, *Metall. Trans. A*, **2A**, 2861–64, 1971.

22. Hondros, E. D. and Seah, M. P., Segregation to interfaces, *Inter. Metals Reviews*, **222**, 262–301, 1977.

23. McLean, D., *Grain Boundaries in Metals*, Clarendon Press, Oxford, 1957.

24. Wynblatt, P. and Ku, R. C., Surface segregation in alloys, in *Interfacial Segregation*, W. C. Johnson and J. M. Blakely (eds), Am. Soc. Metals, Metals Park, OH, 115–36, 1979.

25. Wynblatt, P. and McCune, R. C., Chemical aspects of equilibrium segregation to ceramic interfaces, in *Surfaces and Interfaces in Ceramic and Ceramic-Metal Systems*, Mater. Sci. Research Vol. 14, J. A. Park and A. G. Evans (eds), Plenum Press, New York, 83–95, 1981.

26. Cahn, J. W., The impurity-drag effect in grain boundary motion, *Acta Metall.*, **10**, 789–98, 1962.

27. Lücke, K. and Stüwe, H.-P., On the theory of grain boundary motion, in *Recovery and Recrystallizaion of Metals*, L. Himmel (ed.), Gordon and Breach, New York, 171–210, 1963.

28. Glaeser, A. M., Bowen, H. K and Cannon, R. M., Grain-boundary migration in LiF: I, mobility measurements, *J. Am. Ceram. Soc.*, **69**, 119–26, 1986.

29. Hwang, S. L. and Chen, I.-W., Grain size control of tetragonal zirconia polycrystals using the space charge concept, *J. Am. Ceram. Soc.*, **73**, 3269–77, 1990.

30. Jeong, J.-W., Han, J.-H. and Kim, D.-Y., Effect of electric field on the migration of grain boundaries in alumina, *J. Am. Ceram. Soc.*, **83**, 915–18, 2000.

31. Yan, M. F., Cannon, R. M. and Bowen, H. K., Grain boundary migration in ceramics, in *Ceramic Microstructures '76*, R. M. Fulrath and J. A. Pask (eds), Westview Press, Boulder, Colorado, 276–307, 1977.

32. Bennison, S.-J. and Harmer, M. P., Grain growth kinetics for alumina in the absence of a liquid phase, *J. Am. Ceram. Soc.*, **68**, C22–C24, 1985.

33. Chiang, Y. M. and Kingery, W. D., Grain-boundary migration in nonstoichiometric solid solutions of magnesium aluminate spinel: I, grain growth studies, *J. Am. Ceram. Soc.*, **72**, 271–77, 1989.

34. Rödel, J. and Glaeser, A. M., Anisotropy of grain growth in alumina, *J. Am. Ceram. Soc.*, **73**, 3292–301, 1990.

35. Tsurekawa, S., Ueda, T., Ichikawa, K., Nakashima, H., Yoshitomi, Y. and Yoshinaga, H., Grain boundary migration in Fe-3%Si bicrystal, *Mater. Sci. Forum*, **204–206**, 221–26, 1996.

36. King, A. H., Diffusion induced grain boundary migration, *Inter. Mater. Rev.*, **32**, 173–89, 1987.

37. Handwerker, C. A., Diffusion-induced grain boundary migration in thin films, in *Diffusion Phenomena in Thin Films and Microelectronic Materials*, D. Gupta and P. S. Ho (eds), Noyes Publications, Park Ridge, NJ, 245–322, 1988.

38. (a) Yoon, D. N., Chemically induced interface migration in solids, *Annu. Rev. Mater. Sci.*, **19**, 43–58, 1989.
    (b) Yoon, D. Y., Theories and observations of chemically induced interface migration, *Inter. Mater. Rev.*, **40**, 149–79, 1995.

39. Yoon, D. N., Cahn, J. W., Handwerker, C. A. and Blendell, J. E., Coherency strain induced migration of liquid films through solids, in *Interface Migration and Control of Microstructure*, C. S. Pande, D. A. Smith, A. H. King and J. Walter (eds), Am. Soc. Metals, Metals Park, OH, 19–31, 1986.

40. Brechet, Y. J. M. and Purdy, G. R., A phenomenological description for chemically induced grain boundary migration, *Acta Metall.*, **37**, 2253–59, 1989.

41. den Broeder, F. J. A., Interface reaction and a special form of grain boundary diffusion in the Cr-W system, *Acta Metall.*, **20**, 319–32, 1972.

42. Hillert, M. and Purdy, G. R., Chemically induced grain boundary migration, *Acta Metall.*, **26**, 333–40, 1978.

43. Yoon, D. N. and Huppmann, W. J., Chemically driven growth of tungsten grains during sintering in liquid nickel, *Acta Metall.*, **27**, 973–77, 1979.

44. Li, C. and Hillert, M., A metallographic study of diffusion-induced grain boundary migration in the Fe-Zn system, *Acta Metall.*, **29**, 1949–60, 1981.

45. Rhee, W. H., Song, Y. D. and Yoon, D. N., A critical test for the coherency strain effect on liquid film and grain boundary migration in Mo-Ni-(Co-Sn) alloy, *Acta Metall.*, **35**, 57–60, 1987.

46. Rhee, W. H. and Yoon, D. N., The grain boundary migration induced by diffusional coherency strain in Mo-Ni alloy, *Acta Metall.*, **37**, 221–28, 1989.

47. Baik, Y.-J. and Yoon, D. N., The effect of curvature on the grain boundary migration induced by diffusional coherency strain in Mo-Ni alloy, *Acta Metall.*, **35**, 2265–71, 1987.

48. Kim, J. J., Song, B. M., Kim, D. Y. and Yoon, D. N., Chemically induced grain-boundary migration and recrystallization in PLZT ceramics, *Am. Ceram. Soc. Bull.*, **65**, 1390–92, 1986.

49. Lee, H.-Y., and Kang, S.-J. L., Chemically induced grain boundary migration and recrystallization in Al_2O_3, *Acta Metall. Mater.*, **30**, 1307–12, 1990.

50. Jeong, J. W., Yoon, D. N. and Kim, D. Y., Chemically induced instability at interfaces of cubic ZrO_2-Y_2O_3 grains in a liquid matrix, *Acta Metall. Mater.*, **39**, 1275–79, 1991.

51. Yoon, K. J. and Kang, S.-J. L., Chemical control of the grain boundary migration of $SrTiO_3$ in the $SrTiO_3$-$BaTiO_3$-$CaTiO_3$ system, *J. Am. Ceram. Soc.*, **76**, 1641–44, 1993.

52. Jeon, J.-H. and Kang, S.-J. L., Effect of sintering atmosphere on interface migration of niobium-doped strontium titanate during infiltration of oxide melts, *J. Am. Ceram. Soc.*, **77**, 1688–90, 1994.

53. Jeon, J.-H. and Kang, S.-J. L., Control of interface migration of melt-infiltrated niobium-doped strontium titanates by solute species and atmosphere, *J. Am. Ceram. Soc.*, **81**, 624–28, 1998.

54. Lee, H. Y., Kim, J.-S. and Kang, S.-J. L., Diffusion induced grain boundary migration and enhanced grain growth in BaTiO$_3$, *Interface Science*, **8**, 223–29, 2000.

55. Tiller, W. A., Jackson, K. A., Rutter, J. W. and Chalmers, B., The redistribution of solute atoms during the solidification of metals, *Acta Metall.*, **1**, 428–37, 1953.

56. Rhee, Y.-W., Lee, H. Y. and Kang, S.-J. L., Diffusion induced grain-boundary migration and mechanical property improvement in Fe-doped alumina, *J. Eu. Ceram. Soc.*, **23**, 1667–74, 2003.

57. Hillert, M., On the driving force for diffusion induced grain boundary migration, *Scripta Metall.*, **17**, 237–40, 1983.

58. Cahn, J. W., On spinodal decomposition in cubic crystals, *Acta Metall.*, **10**, 179–83, 1962.

59. Hilliard, J. E., Spinodal decomposition, in *Phase Transformations*, H. I. Aaronson (ed.), Am. Soc. Metals, Metals Park, OH, 497–560, 1968.

60. Chen, F. S., Dixit, G., Aldykiewicz, A. J. and King, A. H., Bicrystal studies of diffusion-induced grain boundary migration in Cu/Zn, *Metall. Trans. A*, **21A**, 2363–67, 1990.

61. Lee, H.-Y., Kang, S.-J. L. and Yoon, D. Y., The effect of elastic anisotropy on the direction and faceting of chemically induced grain boundary migration in Al$_2$O$_3$, *Acta Metall. Mater.*, **41**, 2497–502, 1993.

62. Lee, H.-Y., Kang, S.-J. L. and Yoon, D. Y., Coherency strain energy and the direction of chemically induced grain boundary migration in Al$_2$O$_3$-Cr$_2$O$_3$ and Al$_2$O$_3$-Fe$_2$O$_3$, *J. Am. Ceram. Soc.*, **77**, 1301–06, 1994.

63. Paek, Y. K., Lee, H.-Y. and Kang, S.-J. L., Direction of chemically induced interface migration in Al$_2$O$_3$-anorthite system, *J. Am. Ceram. Soc.*, **79**, 3029–32, 1996.

64. Chung, Y. H., Shin, M. C. and Yoon, D. Y., The effect of external stress on the discontinuous precipitation in an Al-Zn alloy at high and low temperatures, *Acta Metall. Mater.*, **40**, 2177–84, 1992.

65. Mura, T., *Micromechanics of Defects in Solids*, Martinus Nijhoff Publ., Dordrecht, 1987.

66. Hay, R. S., Coherency strain energy and thermal strain energy of thin films in any crystal system, *Scripta Metall.*, **26**, 535–40, 1992.

67. Lee, H.-Y. and Kang, S.-J. L., A general equation of coherency strain energy and its application, *Z. Metallkd.*, **85**, 426–31, 1994.

68. Rehrig, P. W., Messing, G. L. and Trolier-McKinstry, S., Templated grain growth of barium titanate single crystals, *J. Am. Ceram. Soc.*, **83**, 2654–60, 2000.

69. Huang, Y. and Humphreys, F. J., Subgrain growth and low angle boundary mobility in aluminium crystals of orientation {110} ⟨001⟩, *Acta Mater.*, **48**, 2017–30, 2000.

70. Wang, S.-M., Effect of grain boundary structure on diffusion induced grain boundary migration in BaTiO$_3$, MS Thesis, KAIST, Daejeon, 2003.

71. Yoon, K. J., Yoon, D. N. and Kang, S.-J. L., Chemically induced grain boundary migration in SrTiO$_3$, *Ceram. Inter.*, **16**, 151–55, 1990.

72. Baik, Y.-J. and Yoon, D. N., The discontinuous precipitation of a liquid phase in Mo-Ni induced by diffusional coherency strain, *Acta Metall. Mater.*, **38**, 1525–34, 1990.

73. Baik, Y.-J. and Yoon, D. N., Migration of liquid film and grain boundary in Mo-Ni induced by temperature change, *Acta Metall.*, **33**, 1911–17, 1985.

74. den Broeder, F. J. A., Diffusion-induced grain boundary migration and recrystallization, exemplified by the system Cu-Zn, *Thin Solid Films*, **124**, 135–48, 1985.

75. Mittemeijer, E. T. and Beers, A. M., Recrystallization and interdiffusion in thin bimetallic films, *Thin Solid Films*, **65**, 125–35, 1980.

76. Guan, Z. M., Liu, G. X., Williams, D. B. and Notis, M. R., Diffusion-induced grain boundary migration and associated concentration profiles in a Cu-Zn alloy, *Acta Metall.*, **37**, 519–27, 1989.

77. Matthews, J. W. and Crawford, J. L., Formation of grain boundaries during diffusion between single crystal films of gold and palladium, *Phil. Mag.*, **11**, 977–91, 1965.

78. Chae, K.-W., Hwang, C. S., Kim, D.-Y. and Cho, S. J., Diffusion induced recrystallization of TiC, *Acta Mater.*, **44**, 1793–99, 1996.

79. Paek, Y.-K., Lee, H.-Y., Lee, J.-Y. and Kang, S.-J. L., Interface instability in alumina under chemical inequilibrium, in *Mass and Charge Transport in Ceramics(Ceramic Trans. Vol. 71)*, K. Koumoto, L. M. Sheppard and H. Matsubara (eds), Am. Ceram. Soc. Weterville, OH, 333–44, 1996.

80. Paek, Y.-K., Lee, H.-Y. and Kang, S.-J. L., Diffusion induced recrystallization in alumina, *J. Eu. Ceram. Soc.*, **24**, 613–18, 2004.

81. Suh, J. H., The effect of synthesis processes on the microstructure of Y-Ba-Cu-O superconducting system, PhD Thesis, KAIST, Daejeon, 1992.

82. Ko, J. Y., Migration of intergranular liquid films and formation of core-shell grains in TiC-WC-Ni system, MS Thesis, KAIST, Daejeon, 2003.

83. Kim, M.-S., Fisher, J. G., Lee, H.-Y. and Kang, S.-J. L., Diffusion-induced interface migration and mechanical property improvement in the lead magnesium niobate–lead titanate system, *J. Am. Ceram. Soc.*, **86**, 1988–900, 2003.

84. Kim, J.-S. and Kang, S.-J. L., Grain boundary migration and dielectric properties of semiconducting SrTiO$_3$ in the SrTiO$_3$-BaTiO$_3$-CaTiO$_3$ system, *J. Am. Ceram. Soc.*, **82**, 1196–200, 1999.

85. Koo, S.-Y., Lee, G.-G., Kang, S.-J. L., Nowotny, J. and Sorrell, C., Suppression of liquid film migration and improvement of dielectric properties in Nb-doped SrTiO$_3$, *J. Am. Ceram. Soc.*, **87**, 1483–87, 2004.

86. Rios, P. R., Abnormal grain growth in materials containing particles, *Acta Metall. Mater.*, **42**, 839–43, 1994.

87. Rollett, A. D., Srolovitz, D. J. and Anderson, M. P., Simulation and theory of abnormal grain growth anisotropic grain boundary energies and mobilities, *Acta Metall.*, **37**, 1227–40, 1989.

88. Grest, G. S., Srolovitz, D. J. and Anderson, M. P., Computer simulation of grain growth—IV. Anisotropic grain boundary energies, *Acta Metall.*, **33**, 509–20, 1985.

89. Srolovitz, D. J., Grest, G. S. and Anderson, M. P., Computer simulation of grain growth—V. Abnormal grain growth, *Acta Metall.*, **33**, 2233–47, 1985.

90. Kang, M.-K., Kim, D.-Y. and Hwang, N. M., Ostwald ripening kinetics of angular grains dispersed in a liquid phase by two-dimensional nucleation and abnormal grain growth, *J. Eu. Ceram. Soc.*, **22**, 603–12, 2002.

91. Rohrer, G. S., Rohrer, C. L. and Mullins, W. W., Coarsening of faceted crystals, *J. Am. Ceram. Soc.*, **85**, 675–82, 2002.

92. Bolling, G. F. and Winegard, W. C., Grain growth in zone-refined lead, *Acta Metall.*, **6**, 283–87, 1958.

93. Holmes, E. L. and Winegard, W. C., Grain growth in zone-refined tin, *Acta Metall.*, **7**, 411–14, 1959.

94. Lee, B.-K., Chung, S.-Y. and Kang, S.-J. L., Grain boundary faceting and abnormal grain growth in $BaTiO_3$, *Acta Mater.*, **48**, 1575–80, 2000.

95. Lee, S. B., Yoon, D. Y. and Henry, M. F., Abnormal grain growth and grain boundary faceting in a model Ni-base superalloy, *Acta Mater.*, **48**, 3071–80, 2000.

96. Lee, B.-K., Jung, Y.-I., Kang, S.-J. L. and Nowotny, J., {111} twin formation and abnormal grain growth in $(Ba,Sr)TiO_3$, *J. Am. Ceram. Soc.*, **86**, 155–60, 2003.

97. Chung, S.-Y. and Kang, S.-J. L., Intergranular amorphous films and dislocation-promoted grain growth in $SrTiO_3$, *Acta Mater.*, **51**, 2345–54, 2003.

98. Hong, B. S., Kang, S.-J. L. and Brook, R. J., The effect of powder purity and sintering temperature on the microstructure of sintered Al_2O_3, unpublished work, 1988.

99. Koo, J. B. and Yoon, D. Y., The dependence of normal and abnormal grain growth in silver on annealing temperature and atmosphere, *Metall. Mater. Trans. A*, **32A**, 469–75, 2001.

100. Choi, J. S. and Yoon, D. Y., The temperature dependence of abnormal grain growth and grain boundary faceting in 316L stainless steel, *ISIJ International*, **41**, 478–83, 2001.

101. Babcock, S. E. and Balluffi, R. W., Grain boundary kinetics—I. In situ observations of coupled grain boundary dislocation motion, crystal translation and boundary displacement, *Acta Metall.*, **37**, 2357–65, 1989.

102. Rae, C. M. F. and Smith, D. A., On the mechanisms of grain boundary migration, *Phil. Mag.*, **A41**, 477–92, 1980.

103. Merkle, K. L. and Thompson, L. J., Atomic-scale observation of grain boundary motion, *Mater. Lett.*, **48**, 188–93, 2001.

104. Chung, S.-Y., and Kang, S.-J. L., Effect of dislocations on grain growth in $SrTiO_3$, *J. Am. Ceram. Soc.*, **83**, 2828–32, 2000.

105. Lee, M.-G., Choi, S.-Y. and Kang, S.-J. L., Effect of dislocations on grain boundary mobility in $BaTiO_3$, unpublished work, 2004.

106. Bae, S. I. and Baik, S., Determination of critical concentrations of silica and/or calcia for abnormal grain growth in alumina, *J. Am. Ceram. Soc.*, **76**, 1065–67, 1993.

107. Schreiner, M., Schmitt, Th., Lassner, E. and Lux, B., On the origins of discontinuous grain growth during liquid phase sintering of WC-Co cemented carbides, *Powder Metall. Inter.*, **16**, 180–83, 1984.

108. Kang, S.-J. L. and Han, S.-M., Grain growth in Si_3N_4 based materials, *MRS Bull.*, **20**, 33–37, 1995.

109. Jang, C.-W., Kim, J. S. and Kang, S.-J. L., Effect of sintering atmosphere on grain shape and grain growth in liquid phase sintered silicon carbide, *J. Am. Ceram. Soc.*, **85**, 1281–84, 2002.

110. Park, C. W. and Yoon, D. Y., Abnormal grain growth in alumina with anorthite liquid and the effect of MgO addition, *J. Am. Ceram. Soc.*, **85**, 1585–93, 2002.

111. Herring, C., The use of classical macroscopic concepts in surface-energy problems, in *Structure and Properties of Solid Surface*, R. Gomer and C. S. Smith (eds), University of Chicago Press, Chicago, IL, 5–81, 1952.

112. Cahn, J. W. and Hoffman, D. W., A vector thermodynamics for anisotropic surfaces—II. Curved and faceted surfaces, *Acta Metall.*, **22**, 1205–14, 1974.

113. Kim, D.-Y., Wiederhorn, S. M., Hockey, B. J., Handwerker, C. A. and Blendell, J. E., Stability and surface energies of wetted grain boundaries in aluminum oxide, *J. Am. Ceram. Soc.*, **77**, 444–53, 1994.

114. Park, C. W. and Yoon, D. Y., Effects of SiO_2, CaO, and MgO additions on the grain growth of alumina, *J. Am. Ceram. Soc.*, **83**, 2605–609, 2000.

115. Kwon, O.-S., Hong, S.-H., Lee, J.-H., Chung, U.-J., Kim, D.-Y. and Hwang, N. M., Microstructural evolution during sintering of TiO_2/SiO_2-doped alumina: mechanisms of anisotropic abnormal grain growth, *Acta Mater.*, **50**, 4865–72, 2002.

116. Bae, S. I. and Baik, S., Critical concentration of MgO for the prevention of abnormal grain growth in alumina, *J. Am. Ceram. Soc.*, **77**, 2499–504, 1994.

117. Park, Y. J., Hwang, N. M. and Yoon, D. Y., Abnormal growth of faceted (WC) grains in a (Co) liquid matrix, *Metall. Trans. A*, **27A**, 2809–19, 1996.

118. Burton, W. K., Cabrera, N. and Frank, F. C., The growth of crystals and the equilibrium structure of their surfaces, *Phil. Trans. R. Soc. London*, **A243**, 299–358, 1951.

119. Hirth, J. P. and Pound, G. M., *Condensation and Evaporation*, Pergamon Press, Oxford, 77–148, 1963.

120. Flemings, M. C, *Solidification Processes*, McGraw-Hill, New York, 301–26, 1974.

121. Peteves, S. D. and Abbaschian, R., Growth kinetics of solid-liquid Ga interfaces: Part I. experimental, *Metal. Trans. A*, **22A**, 1259–70, 1991.

122. Kang, S.-J. L., Han, S.-M., Lee, D.-D. and Yoon, D. N., $\alpha' \rightarrow \beta'$ phase transition and grain morphology in Y-Si-Al-O-N system, in *MRS Inter. Meeting on Advanced Materials, Vol. 5*, Materials Research Society, 63–67, 1989.

123. Lee, S.-H., Kim, D.-Y. and Hwang, N. M., Effect of anorthite liquid on the abnormal grain growth of alumina, *J. Eu. Ceram. Soc.*, **22**, 317–21, 2002.

108. Kang, E. T. C. and Hall, S. M. Thin films in H. M. Rosenberg, ed., *Phys. Rev.* **20**, 57–71, 1989.

109. Khang, S. W., Xiao, J. X. and Xing, S. J. T., Electrical shielding analysis and phase effects of liquid phase sintered silicon carbide, *J. Am. Ceram. Soc.*, **84**, 1024–1028.

110. Park, G. W. and Yhou, D. Y., Microstructure and growth of silicate with transient liquid and the phase of MgO solid, *Appl. Phys. Rev. Lett. Soc.*, **25**, 236–47, 1980.

111. Hirano, C., The role of the physical microscopic particle in surface-energy process in the time and integrated material devices, R. Gupta, ed., *J. Am. Soc. Tech. Lett. University of Kansas Press, Columbus*, II, 5–51, 1982.

112. Stin, L. W. and Halloway, D. W., A vapor thermodynamics for wire based materials II: Curved and flat of surfaces, *Acta Metall.*, **22**, 259–64, 1994.

113. Knox, J. C., Westbrook, S. M., Henry, R. L., Howard, G. A. and Warren, J. E., Structural surface content of silicon on a bond films in aluminum oxide, *Acta Geram. Soc.*, **77**, 339–51, 1994.

114. Park, C. N. and Yada, K. T., Films in SiO$_2$, SiO$_2$ and MgO surface on the interface.

PART IV
MICROSTRUCTURE DEVELOPMENT

As described in Chapter 1, the basic phenomena occurring during the sintering of a powder compact are densification and grain growth. Parts II and III described densification without grain growth and grain growth in a fully dense compact, respectively. In Part IV, the microstructural development of a compact during sintering will mainly be discussed by analysing these basic phenomena together. Chapter 11 describes the microstructural development of a powder compact containing small spherical pores. At the final stage of sintering, where the pore size is usually smaller than the grain size, the pore shape, which is determined by the dihedral angle, is not the primary parameter by which microstructural development is analysed. In contrast, for pores of size comparable to or larger than the grain size, pore shape is an important parameter. Depending on the dihedral angle and ratio of pore size to grain size, pores cannot shrink unless a critical condition is satisfied as a result of grain growth. This subject is discussed in Chapter 10.

10

GRAIN BOUNDARY ENERGY AND SINTERING

10.1 THE GRAIN BOUNDARY AS AN ATOM SOURCE

As explained in Section 4.1.4, the grain boundary in polycrystalline materials is not an ideal source and sink of atoms, and a critical driving force is needed if boundaries are to operate as a source. The critical driving force for movement of atoms at the boundary varies considerably with the nature, structure and orientation of the grain boundary. For rough boundaries with random orientations, the critical driving force must be very low. On the other hand, for special (or faceted) boundaries with energy cusps, the critical driving force should be high. When the boundary is faceted, the critical driving force is expected to be much higher than that for rough boundaries. According to a recent investigation on sintering of BaTiO$_3$, the densification almost stopped at the final stage of sintering when the grain boundary structure changed from rough to faceted.[1] This result suggests that grain boundary faceting causes a considerable increase in the critical driving force for atom movement from the grain boundary to the neck. In reality, the grain boundary is not an ideal atom source with no energy barrier for operation but a source needing a critical driving force that varies with the nature of the boundary and particularly its structure. Investigations of this subject are, however, very limited. (See Section 4.1.4.)

The critical driving force may also be increased by finely dispersed second-phase particles.[2] If dislocations at a grain boundary are pinned by second-phase particles, excess energy is needed for atom movement from the boundary. In this case the atom flux J from the grain boundary can be expressed as

$$J = \frac{D}{RT}\left(\frac{1}{L}\right)\left(\frac{\gamma_s}{r} - A\right) \tag{10.1}$$

where L is the diffusion distance and A the stress necessary for the movement of atoms from the dislocation source. A may be expressed as

$$A = \frac{2\mu b}{d} \tag{10.2}$$

where μ is the shear modulus, b the Burgers vector, and d the average interparticle distance. This equation is in fact equivalent to that for dislocation generation from a Frank–Read source.[3]
    Therefore, when

$$d < \frac{2\mu b r}{\gamma_s} = \frac{\mu b x^2}{2a\gamma_s} \tag{10.3}$$

material transport from grain boundary to neck will not occur. This means that sintering will be limited when non-sinterable second-phase particles are finely distributed at the grain boundaries and it is found that the presence of fine second-phase particles considerably reduces the sintering kinetics.[4–6]

10.2 EFFECT OF GRAIN BOUNDARY ENERGY ON PORE SHRINKAGE

In solid state sintering, since the surface is replaced by the grain boundary, the grain boundary energy impedes sintering. In real systems, the grain boundary energy is not zero but has finite values. The non-zero grain boundary energy results in the formation of dihedral angles less than 180° and makes the grain surface adjoining a pore either convex, flat or concave. Since the driving force of pore shrinkage arises from the capillary pressure at the grain surface, it can be positive, zero or negative. Kingery and François[7] calculated the critical number of grains surrounding a pore required to make the grain surface flat for a given dihedral angle, a metastable configuration of the pore. Later, in a calculation of the total interfacial energy (surface and grain boundary energies) of grains surrounding a pore, Lange and Kellett[8] showed that there was a minimum energy configuration at a specific pore to grain size ratio for a given number of grains. This approach of calculating the effect of total interfacial energy on the pore stability is more general than the previous method[7] of observing grain surface curvature and is applicable to both local systems with a limited number of grains as well as to the total system with a large number of grains. Consideration of the local system is useful in analysing the microstructural stability at the initial stage of sintering. For this stage, however, the sintering is not directly affected by the reduction in total interfacial energy but by the neck geometry of the grains. In this regard it would be more realistic to limit our discussion on pore stability to the final stage of sintering.

At the final stage of sintering, the number of grains around an isolated pore is determined by the pore size and the average grain size. Since the shape of the surrounding grains is similar to that of the average grain, the number of surrounding grains itself represents the average grain size. Under this condition the interfacial energy approach gives the same result as that of the surface curvature approach, which is physically simpler for discussion of pore stability.

The surface curvature of the grains around an isolated pore is affected by their number and the dihedral angle between them. Figure 10.1 illustrates schematically the three types of surface curvature associated with various numbers of grains for a two-dimensional system and a dihedral angle of 120°. (The assumption of such a dihedral angle is, of course, impossible in reality.) In this case the ratio of pore radius to surface curvature radius is zero, positive and negative when the numbers of grains is six, less than six and more than six, respectively. When the surface radius is infinite, there is no difference in the chemical potential of atoms at the grain boundary and on the grain surface, and the pore is in a metastable state. For more than six grains the grain surface is convex and the chemical potential of atoms on the grain surface is higher than at the grain boundary. The pore has a tendency to expand. In contrast, for a grain number less than six, the pore tends to shrink. For a two-dimensional structure, the condition of pore shrinkage is expressed as

$$\phi > \left(1 - \frac{2}{n}\right)\pi \qquad (10.4)$$

where ϕ is the dihedral angle and n the number of grains surrounding the pore.

Figure 10.2 shows the calculated relationship between pore stability, dihedral angle and number of surrounding grains in a three-dimensional structure.[7] As the dihedral angle increases, the relative pore to grain size

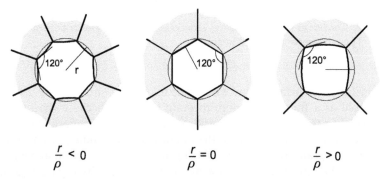

Figure 10.1. Variation of pore shape with the number of surrounding grains. Dihedral angle assumed to be 120°.

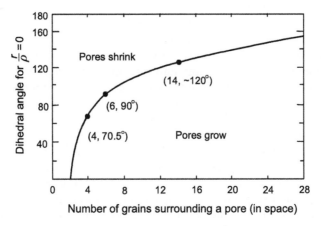

Figure 10.2. Conditions for pore stability: variation of dihedral angle for pore stability with the number of grains surrounding a pore.[7]

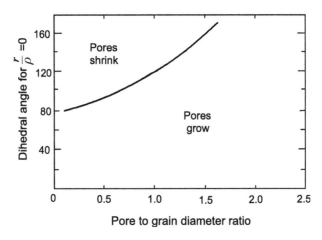

Figure 10.3. Conditions for pore stability: variation of dihedral angle for pore stability with the ratio of pore diameter to grain diameter.[7]

necessary for pore shrinkage increases. The number of surrounding grains can be expressed as a function of pore to grain size ratio because the number is proportional to the pore to grain surface ratio. For tetrakaidecahedral grains to which the nearest number is 14, the number of surrounding grains is 14 for the pore $(2r)$ to grain (G) size ratio of 1. In this case, the dihedral angle may be considered to be 120°. If $2r/G = 2$, the number is $14 \times 4 = 56$. Figure 10.3 shows the variation in dihedral angle with pore to grain size ratio for metastable pores.[7]

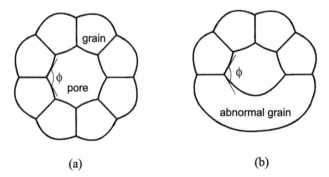

Figure 10.4. Possible benefit of abnormal grain growth for pore shrinkage:[9] (a) pore may expand and (b) pore may shrink.

In real powder compacts where the dihedral angle is, in general, around $150°(\gamma_b \approx (1/3)\gamma_s–(1/2)\gamma_s)$, the pores smaller than the average grain satisfy the shrinkage condition while those larger than a few times the grain size cannot shrink. In real systems, large pores are more frequently observed than small pores during sintering. The above discussion on pore stability suggests that, in addition to pore coalescence, the high stability of large pores can be another cause of the presence of large pores in many sintering compacts. The pore stability concept further emphasizes the importance of the uniform packing of particles in powder compacts for densification. In this regard, possible particle rearrangement during pressure application is an additional benefit of pressure-assisted sintering because it can eliminate any exceptionally large pores in the compact.

According to the concept of pore stability (Figures 10.1–10.3), pores that cannot shrink at the beginning may satisfy their shrinkage condition with grain growth. This means that grain growth can contribute to densification if the compact contains large pores. Xue proposed that even abnormal grain growth could contribute to the shrinkage of large pores (Figure 10.4).[9] A pore which does not satisfy its shrinkage condition (Figure 10.4(a)) can shrink after abnormal grain growth as shown in Figure 10.4(b). However, questions remain as to whether such a contribution of abnormal grain growth is realistic. Nevertheless, it seems clear that, contrary to the classical understanding, grain growth is not always detrimental to densification.

11

GRAIN GROWTH AND DENSIFICATION IN POROUS MATERIALS

As sintering proceeds, both the density and grain size of the compact increase. Since these processes occur simultaneously, interactions between them must be considered if microstructural development during sintering is to be understood. This chapter explains the interaction between densification and grain growth in terms of microstructural development at the final stage of solid state sintering.

11.1 MOBILITY OF AN ISOLATED PORE

At the final stage of sintering, pores are present mostly at grain boundaries and, in particular, at triple junctions. This final stage microstructure suggests that the pores move together with the grain boundaries as the grains grow and it follows that pores at grain boundaries inhibit grain growth.[10] If the pore is considered to be a second phase with a dihedral angle of 180° (see Section 6.2), the maximum inhibiting force F_p^d of a pore with a radius of r against boundary movement, i.e. the force acting on the pore, is $\pi r \gamma_b$. As long as the pore is attached to the grain boundary and both move together, the pore velocity v_p is expressed as

$$v_p = M_p F_p \tag{11.1}$$

where M_p is the pore mobility which depends on the mechanism of pore migration.[11,12]

Pore migration mechanisms include surface diffusion, lattice diffusion, gas diffusion and evaporation/condensation, as shown schematically in Figure 11.1. Under a capillary driving force of grain boundary migration, the atoms in front of the moving boundary around the pore are under compression while those behind the boundary are under tension. This pressure distribution results in a

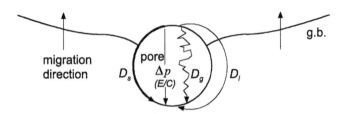

Figure II.I. Possible mechanisms of pore migration with a grain boundary.

difference in the chemical potential of atoms between the two regions and induces atom transport from the compressive region to the tensile region. Let F_p be the force acting on the pore and F_a the force acting on an atom, then

$$F_p \, dx \approx F_a \frac{\pi r^2 dx}{\Omega} \, 2r \qquad (11.2)$$

Here Ω is the atom volume. For pore migration by the surface diffusion of atoms, the material transport rate dV/dt from pore front to pore back is expressed as

$$\frac{dV}{dt} = \pi r^2 \frac{dx}{dt} = JA\Omega$$
$$= N_v \frac{D_s}{kT} F_a A\Omega = \frac{D_s}{kT} 2\pi r \delta_s F_a \qquad (11.3)$$

where N_v is the number of atoms per unit volume and δ_s the thickness of the surface through which diffusion occurs. The rate of pore migration dx/dt is then

$$\frac{dx}{dt} = v_p = \frac{2D_s \delta_s}{kTr} F_a = \frac{2D_s \delta_s}{kTr} \left(\frac{\Omega}{2\pi r^3} F_p \right)$$
$$= \frac{D_s \delta_s \Omega}{\pi r^4 kT} F_p \qquad (11.4)$$

Therefore, the pore mobility by surface diffusion M_p^s is expressed as

$$M_p^s = \frac{D_s \delta_s \Omega}{\pi r^4 kT} = \frac{D_s \delta_s V_m}{\pi r^4 RT} \qquad (11.5)$$

This equation shows that the mobility is inversely proportional to the fourth power of the pore size.

Table II.I. Mobility of pores in porous systems[12]

Migration mechanism	Mobility, M_p
Surface diffusion	$M_p^s = \dfrac{D_s \delta_s \Omega}{\pi r^4 kT} \propto \dfrac{1}{r^4}$
Lattice diffusion	$M_p^l = \dfrac{D_l \Omega}{\pi r^3 kT} \propto \dfrac{1}{r^3}$
Gas diffusion	$M_p^g = \dfrac{D_g p_\infty \Omega^2}{2\pi r^3 (kT)^2} \propto \dfrac{1}{r^3}$
Evaporation/condesation	$M_p^{e/c} = \dfrac{p_\infty \Omega^2}{\sqrt{2mr^2}} \left(\dfrac{1}{\pi kT}\right)^{3/2} \propto \dfrac{1}{r^2}$

Similarly, the pore mobilities by other mechanisms can be calculated and are listed in Table 11.1.[12] For lattice and gas diffusion, the mobility is inversely proportional to the cube of the pore size. In the case of evaporation/condensation, the square of the pore size affects the pore mobility. These mechanisms of pore migration are in fact those of neck growth without shrinkage at the initial stage of sintering. Therefore, the dependence of pore mobility on pore size is the same as that predicted by Herring's scaling law (see Section 4.4.1).[13] In addition, these mechanisms are simultaneously operative in any system. The dominant mechanism can vary with the experimental condition, such as pore size and temperature, as in the case of the sintering diagram. For example, the contribution of surface diffusion increases as the pore size decreases.

II.2 PORE MIGRATION AND GRAIN GROWTH

When pores are present at grain boundaries, the net driving force of grain boundary migration is the difference between the driving force for the boundary without pores and the inhibition force of pores against boundary migration. The boundary velocity v_b is then expressed as

$$v_b = M_b(F_b - NF_p) \tag{11.6}$$

where M_b is the boundary mobility and N the number of pores per unit grain boundary area. For a combined movement of a grain boundary and pores,

$$v_b = v_p = M_p F_p = M_b(F_b - NF_p) \tag{11.7}$$

and hence

$$v_b = \frac{M_b}{1 + N(M_b/M_p)} F_b \tag{11.8}$$

From Eq. (11.8), two extreme cases can be considered[14,15]:

(i) $NM_b \gg M_p$ and
(ii) $NM_b \ll M_p$.

The condition of $NM_b \gg M_p$ applies to a system containing (many) pores with low mobilities. In this case pore migration controls the boundary migration. Then,

$$v_b = \frac{M_p}{N} F_b \equiv M_b^p F_b \tag{11.9}$$

On the other hand, for (a small number of) pores with high mobilities, $NM_b \ll M_p$ is satisfied. The migration of the grain boundary is then controlled by its intrinsic mobility (boundary control). The pores do not affect the boundary migration and v_b is reduced to

$$v_b = M_b F_b \tag{11.10}$$

When M_b equals M_p/N in Eq. (11.8), the grain boundary mobility is the same as the pore mobility (equal mobility).[15,16] The equal mobility condition on a grain size versus pore size plane can be calculated for any mechanism of pore migration. As an example, consider the pore migration controlled by surface diffusion. In this case the pore mobility $M_p = (D_s \delta_s \Omega)/(\pi r^4 kT)$ and the boundary mobility $M_b = D_b^\perp/kT$ for boundaries without impurity segregation. If N is inversely proportional to the boundary area per grain and is expressed as the number of pores per atom at the boundary,[12,15]

$$N \propto \frac{a^2}{G^2} \tag{11.11}$$

where G is the grain size and a^2 the atom area. To satisfy Eq. (11.11), the pore size distribution must be invariable and the number of pores per grain constant during grain growth. When the pores are sufficiently mobile so that they are not entrapped within grains, the pore to grain number ratio is considered to be constant[17] and the above assumption is acceptable. Since N is expressed as

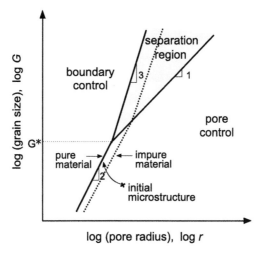

Figure II.2. Dependence of the type of pore/boundary interaction on microstructural parameters (pore size and grain size) when pores move by surface diffusion.[15]

the number of pores per atom, F_b must also be expressed as the force per atom and is

$$F_b = a^2 \Delta P = a^2 \frac{2\gamma_b}{\beta G} \tag{11.12}$$

where β is a constant which is determined by the real curvature of the boundary. Then, from $NM_b = M_p$ for equal mobility,

$$G = r^2 \sqrt{\frac{D_b^{\perp} a^2 \pi}{D_s \delta_s \Omega}} = r^2 \sqrt{\frac{D_b^{\perp} \pi}{D_s \delta_s a}} \tag{11.13}$$

This equation shows that the slope of the equal mobility line is 2 in a log G versus log r plot (Figure 11.2[15]).

II.3 PORE/BOUNDARY SEPARATION

When pores are separated from the grain boundary (pore/boundary separation), the pores are entrapped within grains and cannot be eliminated by sintering or even by hot isostatic pressing.[18] Therefore, pore/boundary separation marks the limit of densification in sintering. Separation occurs when the boundary migration velocity is higher than the pore migration velocity. Since

$$v_p = M_p F_p \tag{11.1}$$

and

$$v_b = v_p = M_p F_p = M_b(F_b - NF_p) \qquad (11.7)$$

for the pores and boundaries to move in combination, pore/boundary separation occurs[12,15,16] when

$$v_p\left(1 + N\frac{M_b}{M_p}\right) < M_b F_b$$

$$F_b > \left(\frac{M_p}{M_b} + N\right)F_p \qquad (11.14)$$

For pore/boundary separation, two extreme cases can be considered:

(i) $M_p < NM_b$ and
(ii) $M_p > NM_b$.

Case (i) corresponds to a system having (a large number of) large pores at the beginning of the final stage of sintering. However, as densification proceeds, the inhibition force of the pores against boundary migration decreases and separation can occur. The separation condition is $F_b > NF_p$ and separation occurs when

$$a^2\frac{2\gamma_b}{\beta G} > \frac{a^2}{G^2}\pi r\gamma_b$$

$$G > \frac{\beta\pi r}{2} \qquad (11.15)$$

This condition corresponds to the Zener condition (see Section 6.2). On the other hand, Case (ii) applies to a system having (a small number of) highly mobile pores and grain boundaries with low velocities. When grain growth occurs and the boundary migration velocity becomes slow, the separated pores can reattach and keep up with the moving boundary. Until this condition is satisfied, the pores are separated. The separation condition is then $F_b > (M_p/M_b)F_p$. For pore migration by surface diffusion, it becomes

$$a^2\frac{2\gamma_b}{\beta G} > \frac{D_s\delta_s\Omega}{\pi r^4 kT}\frac{kT}{D_b^\perp}\pi r\gamma_b$$

$$G < \frac{2a^2}{\beta}\frac{D_b^\perp r^3}{D_s\delta_s\Omega} \qquad (11.16)$$

If the grain boundaries are impure with solute segregation, instead of $D_b^\perp/kT$, $1/\alpha C_\infty$ (Eq. (7.14)) must be inserted for M_b.

So far, the analysis of pore mobility, grain growth and pore/boundary separation has been made under the assumption that the pores are spherical and their shape is unchanged during migration. In reality, however, the pore shape is not spherical and changes as the boundary migrates.[19–22] Evans and co-workers[19,20] examined pore-controlled boundary migration in a theoretical analysis that considered the change in shape of pores and boundaries during migration. In particular, they considered surface-diffusion-controlled pore migration and calculated the pore shape and pore migration velocity as a function of the dihedral angle ϕ.[19] They found that the peak pore velocity, v_p, in a steady state is expressed as

$$
\begin{aligned}
v_p &\approx \frac{\gamma_s D_s \delta_s \Omega}{kTr^3}(17.9 - 6.2\phi) \\
&\equiv \frac{\gamma_b D_s \delta_s \Omega}{kTr^3}\frac{(17.9 - 6.2\phi)}{2\cos(\phi/2)}
\end{aligned}
\tag{11.17}
$$

Later, Svoboda and Riedel[21,22] studied more rigorously the pore shape, pore migration and grain coarsening for various types of pores. The predicted shape change of a flat pore was consistent with the experimental observation of Rödel and Glaeser.[23,24] The velocity of an isolated pore was also derived to be similar to Eq. (11.17) except for a numerical constant which depends on the dihedral angle.

Equation (11.17) indicates that the peak pore velocity increases as the dihedral angle decreases and can exceed the value[15] anticipated by the phenomenological model with a spherical pore by up to an order of magnitude. On the other hand, as the dihedral angle approaches 180°, the peak velocity decreases drastically and can have a negative value. This result means that pores with a dihedral angle close to 180° cannot move together with grain boundaries in a steady state and will always detach from them. This consequence would be reasonable because there can be no preference for spherical pores with $\phi = 180°$ to locate on grain boundaries with zero energy. The peak velocity of an actual pore with $\phi < 180°$, however, is the same as that of the phenomenological model except for a numerical constant. The physical meanings of the models are therefore the same.

Figure 11.2[15] delineates the calculated conditions of pore/boundary separation (Eqs (11.15) and (11.16)) together with the equal mobility line on a grain size versus pore size plane in log scale when pores move by surface diffusion. The minimum grain size G^* for pore/boundary separation is determined to be the point where the two lines of pore/boundary separation conditions converge. For surface-diffusion controlled boundary movement, G^* is expressed as

$$
G^* = \sqrt{\frac{\beta^4 \pi^3 D_s \delta_s \Omega}{16 D_b^\perp a^2}}
\tag{11.18}
$$

It appears that G^* is proportional to the square root of the pore mobility and inversely proportional to the square root of the boundary mobility. To reduce the chance of pore/boundary separation, densification relative to grain growth should be improved or G^* should be increased. Improvement of densification means a raise in D_l or $D_b^{\parallel}$, and a reduction in D_s. On the other hand, an increase in G^* is achievable by reducing $D_b^{\perp}$ and enhancing D_s. The modification of diffusion coefficient is achievable by changing temperature or adding appropriate dopants. In particular, the addition of dopants with high grain boundary segregation can be useful in reducing $D_b^{\perp}$ without loss of sinterability (Figure 11.2). However, secondary effects of dopants, such as changes in grain boundary and surface energy, and also their anisotropy may dominate the overall kinetics. Comparison of Eqs (11.18) and (11.5) shows that low D_s is preferred at the beginning of sintering to suppress grain growth while high D_s, a contradictory requirement, is desirable at the final stage of sintering to reduce the chance of pore/boundary separation.[25,26] Two such opposite requirements for the same variable may also signify the complexity and difficulty of sintering.

The microstructural development during sintering can be represented by a trajectory from the initial microstructure on a log G versus log r plane, as in Figure 11.2. Such a trajectory gives information on pore and grain size change during sintering. However, since sintered density has a more practical meaning than pore size, microstructural development is better represented on a grain size versus density plane, as in Figure 11.3.[25-27] In this case the pore/boundary

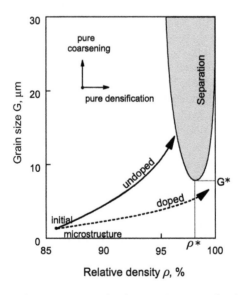

Figure II.3. Schematic of microstructure development in terms of a plot of grain size versus density.[25-27]

separation line determined by the Zener condition appears on the left-hand side of a separation region, and the direction of the densification trajectory is opposite to that in Figure 11.2.

In addition to theoretical studies on microstructure development (Figures 11.2 and 11.3) based on pore mobility, pore-controlled grain growth and pore/boundary separation, experimental studies have also been made.[23,24,28] In particular, using a photolithographic technique, Rödel and Glaeser[23,24] made model pores of controlled size and shape between Al_2O_3 single crystals and polycrystals and studied the pore mobility and pore/boundary separation. Their study confirmed experimentally the theoretical analysis and understanding of microstructural development during sintering.

11.4 MICROSTRUCTURE DEVELOPMENT IN A POROUS COMPACT

The microstructural development at the final stage of sintering is characterized by densification and pore-dragged grain growth. Therefore, it is determined by the relative densification rate $((1/\rho)(d\rho/dt))$ and the relative grain growth rate $((1/G)(dG/dt))$, which are affected by densification and grain growth mechanisms.[25–28] Under the condition that pore movement controls boundary movement, densification occurs by lattice or grain boundary diffusion and grain growth by surface diffusion, gas phase transport or lattice diffusion. Microstructure development can then be categorized into three different types in terms of the dependency on grain size. Let m be the exponent of grain size for densification and n that for grain growth. Then, the three cases are

(i) $m < n$,
(ii) $m = n$ and
(iii) $m > n$,

as in the following examples.

11.4.1 Case I: $m < n$

Example: densification by lattice diffusion and grain growth by surface diffusion.

Densification*

$$\frac{1}{\rho}\frac{d\rho}{dt} \propto \frac{D_l \gamma_s V_m (1-\rho)^{1/3}}{RTG^3 \rho} \tag{11.19}$$

---

*Here, Eq. (5.9) is taken instead of Eq. (5.8) (Coble's equation) for densification by lattice diffusion.

Grain growth

$$v_b = \frac{dG}{dt} = \frac{M_p}{N} F_b$$

$$= \frac{D_s \delta_s V_m}{\pi r^4 RT} \frac{G^2}{a^2} \frac{2\gamma_b a^2}{\beta G} = \frac{2\gamma_b D_s \delta_s V_m G^2}{\beta \pi r^4 RTG}$$

$$r^3 \propto \text{porosity} \times G^3 = (1-\rho)G^3$$

$$\therefore \frac{1}{G} \frac{dG}{dt} \propto \frac{D_s \delta_s \gamma_b V_m}{RTG^4 (1-\rho)^{4/3}} \tag{11.20}$$

Figure 11.4(a) depicts the relative densification and grain growth rates with grain size. At small grain sizes, grain growth dominates densification. This is evident on the basis of the grain size dependency in Eqs (11.19) and (11.20) where $m < n$.[†]

Therefore, the advantage of using a fine powder for densification is reduced because of fast grain growth. On the other hand, since densification time increases in proportion to the cube of powder size (G^3), the use of coarse powder is also undesirable. In this regard it is useful to increase the relative densification rate and reduce G^*, as shown schematically in Figure 11.4(b).

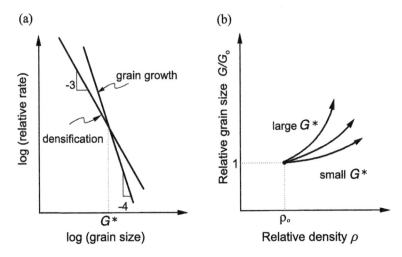

Figure II.4. (a) Relative densification and coarsening rates versus grain size and (b) microstructure development for the case where densification is controlled by lattice diffusion and grain growth by surface diffusion.[26,27]

[†]The simple comparison of log *rate* versus log G is only for a given relative density because G and ρ are interrelated. The conventional scaling law must be modified for interrelated densification and grain growth (see Section 11.5).

To reduce $G^*[\propto (D_s\delta_s\gamma_b/D_l\gamma_s)]$, D_s/D_l and γ_b/γ_s must be reduced. With the reduction of G^*, however, the region of pore/boundary separation is enlarged and pore/boundary separation can easily occur. An optimization is needed as always.

II.4.2 Case II: $m = n$

Example: densification by grain boundary diffusion and grain growth by surface diffusion.

Densification

$$\frac{1}{\rho}\frac{d\rho}{dt} \propto \frac{D_b\delta_b\gamma_s V_m}{RTG^4\rho} \tag{11.21}$$

Grain growth

$$\frac{1}{G}\frac{dG}{dt} \propto \frac{D_s\delta_s\gamma_b V_m}{RTG^4(1-\rho)^{4/3}} \tag{11.20}$$

In this case, since $m = n$, the ratio of relative growth rate to relative densification rate $[\Gamma = (1/G)(dG/dt)/(1/\rho)(d\rho/dt)]$ is independent of grain size (Figure 11.5(a)). Figure 11.5(b) depicts schematically the sintering trajectories for small and large ratios of relative rates (Γ).

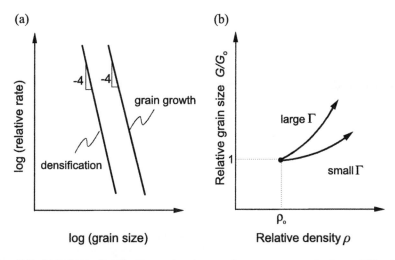

Figure II.5. (a) Relative densification and grain growth rates versus grain size and (b) microstructure development for the case where densification is controlled by grain boundary diffusion and grain growth by surface diffusion. Γ is the ratio of relative growth rate to relative densification rate.

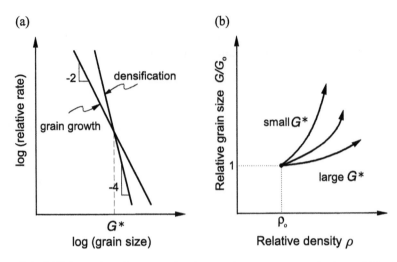

Figure II.6. (a) Relative densification and grain growth rates versus grain size and (b) microstructure development for the case where densification is controlled by grain boundary diffusion and grain growth by evaporation/condensation.

II.4.3 Case III: $m > n$

Example: densification by grain boundary diffusion and grain growth by evaporation/condensation.

Densification

$$\frac{1}{\rho}\frac{d\rho}{dt} \propto \frac{D_b \delta_b \gamma_s V_m}{RTG^4 \rho} \tag{11.21}$$

Grain growth

$$\frac{1}{G}\frac{dG}{dt} \propto \frac{p_\infty \Omega^2}{\sqrt{m}}\left(\frac{1}{\pi kT}\right)^{3/2}\frac{\gamma_b}{G^2(1-\rho)^{2/3}} \tag{11.22}$$

As $m > n$, grain growth dominates at large particle sizes (Figure 11.6(a)). Figure 11.6(b) is just the opposite of Figure 11.4(b). For the sintering of materials with a high vapour pressure at sintering temperatures, for example Si_3N_4 and SiC with a covalent bond nature, the use of fine powder has benefits in obtaining a dense and fine microstructure after sintering.

II.5 SCALING LAW AT FINAL STAGE SINTERING

The effect of particle (or grain) size on microstructure at the final stage of sintering is not as simple as that at the initial stage. At the initial stage where no grain growth is assumed, the neck growth and sample shrinkage depend only on initial particle size and their relationship follows Herring's scaling law[13] (see Section 4.4.1). On the other hand, since grain growth and densification occur simultaneously at the final stage, not only the initial grain size but also the grain growth kinetics affect the densification. Therefore, for two sintering compacts with similar microstructures having the same number and location of pores and the same pore to grain size ratio, the period of time to achieve the same density or grain size can be different depending on the mechanism prevailing for the other phenomenon. This is because the dependencies of densification and growth on grain size are, in general, different.

To establish the scaling laws of densification and growth at the final stage of sintering, the interaction between the two phenomena has therefore to be considered. For simplicity, we may make the following assumptions:

(i) the mechanisms of densification and growth are unchanged during sintering, as in the case of Herring's scaling law;
(ii) the number and type of pores per grain are constant, following experimental observations that the pore number and type are almost unchanged if pores are not entrapped within grains,[17] and
(iii) no pore/boundary separation occurs.

For a constant number of pores per grain, the densification and grain growth equations take the form (Table 11.2)

$$\frac{1}{\rho}\frac{d\rho}{dt} = \frac{K_1(1-\rho)^k}{G^m\rho} \tag{11.23}$$

and

$$\frac{1}{G}\frac{dG}{dt} = \frac{K_2}{G^n(1-\rho)^l} \tag{11.24}$$

where K_1 and K_2 are constants containing various parameters, for example, diffusivity, surface energy, temperature and molar volume, and k, l, m and n are exponents. Equation (11.23) is from Eq. (5.9) of lattice diffusion and Eq. (5.10) of grain boundary diffusion. Combining Eq. (11.23) and Eq. (11.24),

$$\frac{d\rho}{dG} = \left(\frac{K_1}{K_2}\right)G^{n-m-1}(1-\rho)^{k+l} \tag{11.25}$$

When sintering proceeds from the initial density of ρ_0 and the initial grain size of G_0 to a density of ρ and a grain size of G, the following equation

$$\int_{\rho_0}^{\rho}\frac{d\rho}{(1-\rho)^{k+l}} = \left(\frac{K_1}{K_2}\right)\int_{G_0}^{G}G^{n-m-1}dG \tag{11.26}$$

Table II.2. Rate equations of densification and grain growth during final stage sintering

Relative densification rate

$$\left(\frac{1}{\rho}\frac{d\rho}{dt}\right) = \frac{K_1(1-\rho)^k}{G^m\rho}$$

Lattice diffusion	Grain boundary diffusion
$\dfrac{K_{11}D_l\gamma_s V_m(1-\rho)^{1/3}}{RTG^3\rho}$	$\dfrac{K_{12}D_b\delta_b\gamma_s V_m}{RTG^4\rho}$

Relative grain growth rate

$$\left(\frac{1}{G}\frac{dG}{dt}\right) = \frac{K_2}{G^n(1-\rho)^l}$$

Surface diffusion	Gas phase diffusion	Evaporation/condensation	Boundary mobility
$\dfrac{K_{21}D_s\delta_s\gamma_b V_m}{RTG^4\rho(1-\rho)^{4/3}}$	$\dfrac{K_{22}D_g p_\infty \gamma_b V_m^2}{(RT)^2 G^3(1-\rho)}$	$\dfrac{K_{23}p_\infty\Omega^2}{\sqrt{m}}\left(\dfrac{1}{\pi kT}\right)^{3/2}\dfrac{\gamma_b}{G^2(1-\rho)^{2/3}}$	$\dfrac{K_{24}D_b^\perp\gamma_b V_m}{\omega RTG^2}$

satisfies. The analytical solution of this equation is a $G - \rho$ trajectory, as shown schematically in Figure 11.3, which shows the nature of microstructural development during sintering. (See Section 11.4.) When the analytical solution of Eq. (11.26) is inserted in Eqs (11.23) and (11.24), $d\rho/dt$ can be expressed only as a function of ρ and dG/dt only as a function of G. Therefore, we can quantitatively predict the effect of grain size on densification and grain growth.

Depending on the grain size exponents of densification (m) and grain growth (n) in Table 11.2, the interrelationship can be categorized into three different types, as in Section 11.4:

  (i) $m = n$,
 (ii) $m < n$ and
(iii) $m > n$.

11.5.1 Case I: $m = n$

This case applies to systems with the same grain size exponent for densification and grain growth, as in densification by grain boundary diffusion and growth by surface diffusion ($m = n = 4$), or densification by lattice diffusion and growth by gas diffusion ($m = n = 3$). Since the effect of grain size on relative densification is the same as that on relative growth, the grain size dependency is the same for densification and growth. When particle size increases λ times, the time needed for the same degree of densification increases by $\lambda^m (= \lambda^n)$ and the time needed for the same degree of grain growth by $\lambda^n (= \lambda^m)$.

11.5.2 Cases II and III: $m \neq n$ ($m < n$ and $m > n$)

Examples of systems for $m \neq n$ are those with densification by lattice diffusion and growth by surface diffusion ($m < n$), and densification by lattice diffusion and growth by boundary diffusion ($D_b^{\perp}$)($m > n$). Since the effects of grain size on densification and growth are different, size effects are not as simple as those in Case I. The scale exponent α in the scaling laws for densification and growth can be calculated using Eqs (11.23), (11.24) and (11.26); but α varies considerably with not only the initial and final conditions but also various physical parameters. The example in Figure 11.7 shows the variation of α for densification with system parameters during sintering with lattice-diffusion-controlled densification and surface-diffusion-controlled growth. Depending on the value of $[(D_l\gamma_s G_o/D_s\delta_s\gamma_b)]$, α has a value between zero and 3. However, $\alpha = 0$ is improbable in reality because grain growth highly dominates densification and essentially no densification results under this condition. For the same system the growth exponent has a value slightly larger than 4 irrespective of the parameters. For a system with densification by lattice diffusion and growth by evaporation/condensation, the densification exponent is also between zero and 3. However, the growth exponent is a

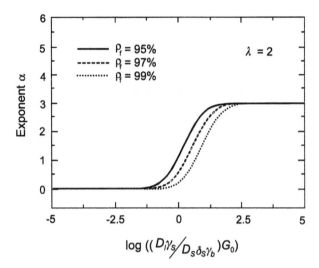

Figure II.7. Variation of densification exponent α with system constants at final stage sintering. Densification is assumed to occur from 90% relative density to ρ_f by lattice diffusion and grain growth by surface diffusion.

constant ($\sim$2) as in the case of surface-diffusion-controlled growth. As the exponent for densification varies with the initial and final conditions, and system parameters, the scaling law for the final stage of sintering cannot be as simple as that of Herring. For predicting the effect of grain size on densification in real systems, experimental conditions must be taken into account in using Eqs (11.23), (11.24) and (11.26).

II.6 MODIFICATION OF THERMAL CYCLE AND MICROSTRUCTURE DEVELOPMENT

Modification of the thermal cycle provides opportunities to improve the sinterability of powder compacts. Typical examples are fast firing and heating-rate-controlled sintering.

II.6.1 Fast Firing

Fast firing was proposed by Brook et al.[29,30] as a technique to suppress grain growth and enhance densification. Figure 11.8 shows schematically the three different thermal cycles used in conventional sintering, fast firing and rate-controlled sintering which will be explained in Section 11.6.2. Fast firing, a sintering technique with a heating rate much faster than that in conventional sintering, can be utilized to sinter any compact where the activation energy of densification, Q_d, is higher than that of grain growth, Q_g (Figure 11.9). When Q_d is larger than Q_g, the ratio of densification rate to grain growth rate

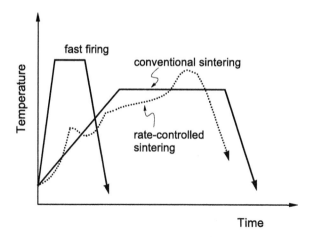

Figure II.8. Schematic showing the thermal cycle of conventional sintering, fast firing, and rate-controlled sintering.

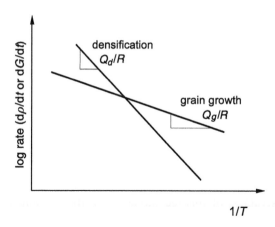

Figure II.9. Temperature dependence of densification and grain growth for a material in which $Q_d > Q_g$.[29]

increases as temperature increases. Fast heating of a compact, then, allows minimization of grain coarsening during heating. Sintering at a temperature higher than that of conventional sintering also enhances the densification rate relative to grain growth rate.

The applicability of fast firing has been confirmed in several systems, such as Al_2O_3[29], $BaTiO_3$[30], Al_2O_3-TiC[31] and ZrO_2[32]. In particular, pressureless sintering of Al_2O_3-TiC composites that had usually been prepared by hot pressing became possible by fast firing processes that allowed fast densification of the compact and suppression of TiC decomposition.[31] In fast firing, the thermal shock due to a fast heating rate, and incomplete elimination of binder and adsorbed impurities must be minimized.[32] To achieve this, a long holding time of powder compacts at low temperatures and fast heating through a temperature range of less than a few hundred degrees Celsius to a fast firing temperature is needed. The microstructure obtained by fast firing is finer than that by conventional sintering but it is sometimes non-uniform due to uneven densification. In particular, for liquid phase sintered compacts with a small volume fraction of liquid, the tendency of liquid coagulation in the centre of the compacts increases and this non-uniform microstructure can remain for a long sintering time.[33]

II.6.2 Heating-Rate-Controlled Sintering

Rate-controlled sintering by control of heating rate and temperature according to predetermined densification rates was proposed by Huckabee and Palmour III.[34] This sintering technique requires prior determination of the desired densification rates during heating. The predetermined densification rates are then achieved by automatic control of the heating rate. A thermal cycle of rate-controlled sintering is, therefore, very different to that of conventional sintering with a constant heating rate, as schematically shown in Figure 11.8, and varies with both the system concerned and the raw materials for the same system.

Unlike conventional sintering with a constant heating rate, rate-controlled sintering involves changes in the heating cycle in an attempt to meet a micro-structural change during heating. Huckabee and Palmour III[34] claimed that their rate-controlled sintering had allowed the fabrication of MgO-doped Al_2O_3 compacts with a finer grain size than that of compacts prepared by conventional sintering. However, the claimed advantage of finer microstructure at the same density has rarely been reported in other systems. Probably, the practical usefulness of this technique lies in the possibility of minimizing potential problems that might occur during the heating-up stage. It would be possible, for example, to eliminate adsorbed gas or volatile materials at an optimum condition.

PROBLEMS

4.1. For any real system where the dihedral angle ϕ is smaller than 180°, there exists a critical size of isolated pores relative to grain size that cannot shrink by pressureless sintering (atmospheric pressure sintering). This result can be explained in terms of either the curvature of the isolated pore or the total interfacial energy. Is there a fundamental difference between the two explanations? Discuss.

4.2. What does a high dihedral angle mean in the initial and final stage of sintering of real powder compacts?

4.3. In a polycrystal, the stability of a pore is governed by the relative size of the pore to the grain size and the relative ratio of surface energy to grain boundary energy. In a glass with no grain boundaries, can a pore be stable? Discuss.

4.4. Consider isolated pores that move along grain boundaries by lattice and surface diffusion of atoms. Draw a schematic figure that shows the variation in pore migration velocity with pore size, and explain. Assume that the driving force of grain boundary migration is constant.

4.5. Derive equations of pore migration by (a) gas diffusion and (b) evaporation/condensation.

4.6. For pore migration (movement) by gas diffusion, what is the effect of pore size on pore migration velocity? The gas pressure in the pore is assumed to be in equilibrium with the pore capillary pressure.

4.7. When a pore at a grain boundary moves by the evaporation/condensation mechanism, the pore mobility is inversely proportional to the square of the pore radius. What is the velocity ratio of pores with a radius of r_1 and a radius of r_2 in two different samples with the same porosity? Assume that the driving force of grain growth is the same for the two samples.

4.8. (a) What is the maximum drag force of a spherical pore against grain boundary movement? Assume a non-zero and constant grain boundary energy.

   (b) If the dihedral angle decreases by a reduction of the solid/vapour interfacial energy, how does the drag force of the pore change with the dihedral angle? Pore volume is assumed to be invariable.

   (c) For pore migration by surface diffusion, explain the change in the pore migration rate (velocity) when the dihedral angle is reduced. Assume a constant driving force for grain boundary migration.

4.9. If grain growth is affected by gas diffusion in pores with a constant pressure, what is the activation energy of grain growth?

4.10. Describe possible techniques (as many as possible) to suppress pore/boundary separation during sintering.

4.11. Consider a system where densification occurs by lattice diffusion and grain growth by surface diffusion. How can you determine the optimum size of the starting powder in view of sintering and powder production cost? If you adopt hot pressing to produce sintered compacts, do you expect the optimum powder size to change? If so, how?

4.12. Describe the basic assumptions involved in the microstructure development map in Figure 11.2 and discuss their validity. To enhance densification while suppressing grain growth at the beginning of the final stage of sintering, what conditions should be provided?

4.13. Draw log G versus log r diagrams at the final stage of sintering and explain for the following cases:
  (a) densification by grain boundary diffusion and grain growth by gas diffusion, and
  (b) densification by grain boundary diffusion and grain growth by surface diffusion.

4.14. At the final stage of sintering of a powder compact, the densification occurs by lattice diffusion and the grain growth by surface diffusion. When you raise the sintering temperature from T_1 to T_2:
  (a) how will the sintered density versus grain size trajectory in Figure P4.14 change?
  (b) how about the pore/boundary separation region?
  (c) will the critical grain size where the relative densification rate is equal to the relative grain growth rate change?

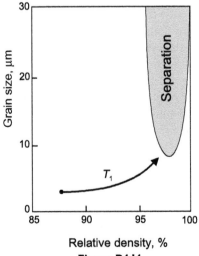

Figure P4.14

4.15. Consider a system where densification occurs by lattice diffusion and grain growth by gas diffusion. Explain the effect of sintering temperature on relative densification and grain growth rates. Assume that the gas pressure in the pores is invariable with temperature.

4.16. Figure P4.16 shows the microstructural development trajectory of an Al_2O_3 powder compact with an initial particle size of $5\,\mu m$.

 (a) Show and explain the microstructural development trajectory of another Al_2O_3 powder compact with an initial particle size of $0.5\,\mu m$. Assume that the densification occurs by grain boundary diffusion (D_b) and the grain growth by surface diffusion (D_s).

 (b) If a dopant addition enhances D_s by ten times, what will be the trajectories of compacts with $5\,\mu m$ and $0.5\,\mu m$ particle sizes?

 (c) If densification occurs by D_b and grain growth by solute drag of the grain boundary, what will the trajectory of the compact with $0.5\,\mu m$ particle size be? Compare the result with that of (a).

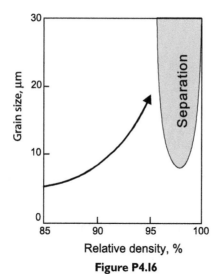

Figure P4.16

4.17. Draw schematically the $G-\rho$ trajectories and explain for two powder compacts (a) with different green densities (low and high) and (b) with the same green density but with different pore size distributions (narrow and broad). Assume that the compacts were made from the same starting powder and that densification occurs by lattice diffusion and grain growth by surface diffusion.

4.18. Consider the grain boundary between two spherical particles with roughly the same radius. In one case, both of the particles are single crystalline. In the other case, one particle is polycrystalline while the other is single crystalline. Compare the neck growth and grain growth for the two cases.

4.19. Discuss whether Herring's scaling law is applicable to a system where densification occurs by lattice diffusion and grain growth by surface diffusion. Assume that all the pores are at the grain boundary and that the number of pores per grain is invariable during sintering.

4.20. Consider a system where densification occurs by lattice diffusion and grain growth by surface diffusion.
   (a) Describe how the activation energy of diffusion can be obtained from diffusivity data.
   (b) Determine a desirable sintering cycle (T vs. t) to enhance densification while minimizing grain growth.

4.21. At the final stage of sintering of alumina powder compacts, densification and grain growth are reported to follow the equations

$$\frac{d\rho}{dt} = \frac{733 D_b \delta_b \gamma_s V_m}{G^4 RT}$$

and

$$\frac{dG}{dt} = \frac{110 D_s \delta_s \gamma_b V_m}{G^3 (1-\rho)^{4/3} RT}$$

respectively. Draw a microstructural development map (G vs. ρ) for an alumina compact with a grain size of $1\,\mu m$ and a relative density of 0.9 during sintering at 1500°C. Note $D_b \delta_b = 8.6 \times 10^{-10} \exp(-418 kJ/RT) m^3/s$, $D_s \delta_s = 1.26 \times 10^{-7} \exp(-493 kJ/RT) m^3/s$, $\gamma_s = 0.71\,J/m^2$, $\gamma_b = 0.34\,J/m^2$ and $V_m = 2.56 \times 10^{-5}\,m^3$.[35–38]

4.22. Describe possible processes for preparing sintered compacts with the following microstructures:
   (a) fully densified compacts with different grain sizes (at least 10 times difference in size);
   (b) compacts with the same grain and pore sizes but with different porosities, say 1–5 vol%;
   (c) compacts with the same grain size and porosity but with different pore sizes, say 1–10 μm.

4.23. Consider powder compacts consisting of two kinds of powders with very low (A) and very high (B) sinterabilities. What do you expect the change in densification rate to be with decreasing powder sizes for the two kinds of powder compacts:
   (a) with a high volume fraction, say over 90%, of powder A and a low volume fraction of powder B, and
   (b) a high volume fraction of powder B and a low volume fraction of powder A.

4.24. For some systems, as a powder compact reaches a sintering temperature, the sintered density shows approximately the same value irrespective of heating rate, the so-called point density phenomenon.[39] Explain possible causes of this phenomenon.

REFERENCES

1. Choi, S.-Y. and Kang, S.-J. L., Sintering kinetics by structural transition at grain boundaries in barium titanate, *Acta Mater.*, **52**, 2937–43, 2004.
2. Ashby, M. F., On interface-reaction control of Nabarro-Herring creep and sintering, *Scripta Metall.*, **3**, 837–42, 1969.
3. Barrett, C. R., Nix, W. D. and Tetelman, A. S., *The Principles of Engineering Materials*, Prentice-Hall, Englewood Cliffs, New Jersey, 240–46, 1973.
4. Early, J. G., Lenel, F. V. and Ansell, G. S., The material transport mechanism during sintering of copper-powder compacts at high temperatures, *Trans. AIME*, **230**, 1641–50, 1964.
5. Brett, J. and Seigle, L., The role of diffusion versus particle flow in the sintering of model compacts, *Acta Metall.*, **14**, 575–82, 1966.
6. Maekawa, K., Nakada, Y. and Kimura, T., Origins of hindrance in densification of Ag/Al$_2$O$_3$ composites, *J. Mater. Sci.*, **37**, 397–410, 2002.
7. Kingery, W. D. and François, B., The sintering of crystalline oxides, I. Interactions between grain boundaries and pores, in *Sintering and Related Phenomena*, G. C. Kuczynski, N. A. Hooton and C. F. Gibbon (eds), Gordon and Breach, New York, 471–98, 1967.
8. Lange, F. F. and Kellett, B., Influence of particle arrangement on sintering, in *Science of Ceramic Chemical Processing*, L. L. Hench and D. R. Ulrich (eds), Wiley, New York, 561–74, 1986.
9. Xue, L. A., Thermodynamic benefit of abnormal grain growth in pore elimination during sintering, *J. Am. Ceram. Soc.*, **72**, 1536–37, 1989.
10. Kingery, W. D. and François, B., Grain growth in porous compacts, *J. Am. Ceram. Soc.*, **48**, 546–47, 1965.
11. Shewmon, P. G., The movement of small inclusions in solids by a temperature gradient, *Trans. Metall. Soc. AIME*, **230**, 1134–37, 1964.
12. Brook, R. J., Controlled grain growth, in *Ceramic Fabrication Processes*, F. F. Y. Wang (ed.), Academic Press, New York, 331–64, 1976.
13. Herring, C., Effect of change of scale on sintering phenomena, *J. Appl. Phys.*, **21**, 301–303, 1950.
14. Nichols, F. A., Further comments on the theory of grain growth in porous compacts, *J. Am. Ceram. Soc.*, **51**, 468–69, 1968.
15. Brook, R. J., Pore-grain boundary interactions and grain growth, *J. Am. Ceram. Soc.*, **52**, 56–57, 1969.
16. Carpay, F. M. A., The effect of pore drag on ceramic microstructures, in *Ceramic Microstructure '76*, R. M. Fulrath and J. A. Pask (eds), Westview Press, Boulder, Colorado, 261–75, 1977.

17. Thompson, A. M. and Harmer, M. P., Influence of atmosphere on final-stage sintering kinetics of ultra-high-purity alumina, *J. Am. Ceram. Soc.*, **76**, 2248–56, 1993.

18. Kwon, S.-T., Kim, D.-Y., Kang, T.-K. and Yoon, D. N., Effect of sintering temperature on the densification of Al$_2$O$_3$, *J. Am. Ceram. Soc.*, **70**, C69–70, 1987.

19. Hsuch, C. H., Evans, A. G. and Coble, R. L., Microstructure development during final/intermediate stage sintering I. pore/grain boundary separation, *Acta Metall.*, **30**, 1269–79, 1982.

20. Spears, M. A. and Evans, A. G., Microstructure development during final/intermediate stage sintering II. grain and pore coarsening, *Acta Metall.*, **30**, 1281–89, 1982.

21. Svoboda, J. and Riedel, H., Pore-boundary interactions and evolution equations for the porosity and the grain size during sintering, *Acta Metall. Mater.*, **40**, 2829–40, 1992.

22. Riedel, H. and Svoboda, J., A theoretical study of grain growth in porous solids during sintering, *Acta Metall. Mater.*, **41**, 1929–36, 1993.

23. Rödel, J. and Glaeser, A. M., Pore drag and pore-boundary separation in alumina, *J. Am. Ceram. Soc.*, **73**, 3302–12, 1990.

24. Glaeser, A. M., The role of interfaces in sintering: an experimental perspective, in *Science of Ceramic Interfaces*, J. Nowotny (ed.), Elsevier Science Publishing, New York, 287–322, 1991.

25. Yan, M. F., Microstructural control in the processing of electronic ceramics, *Mater. Sci. Eng.*, **48**, 53–72, 1981.

26. Brook, R. J., Fabrication principles for the production of ceramics with superior mechanical properties, *Proc. Brit. Ceram. Soc.*, **32**, 7–24, 1982.

27. Harmer, M. P., Use of solid-solution additives in ceramic processing, in *Structure and Properties of MgO and Al$_2$O$_3$ Ceramics*, W. D. Kingery (ed.), Am. Ceram. Soc. Inc., Columbus, Ohio, 679–96, 1985.

28. Handwerker, C. A., Cannon, R. M. and Coble, R. L., Final-stage sintering of MgO, in *Structure and Properties of MgO and Al$_2$O$_3$ Ceramics*, W. D. Kingery (ed.), Am. Ceram. Soc. Inc., Columbus, Ohio, 619–43, 1985.

29. Harmer, M. P. and Brook, R. J., Fast firing—microstructural benefits, *J. Br. Ceram. Soc.*, **80**, 147–48, 1981.

30. Mostaghaci, H. and Brook, R. J., Production of dense and fine grain size BaTiO$_3$ by fast firing, *Trans. J. Br. Ceram. Soc.*, **82**, 167–70, 1983.

31. Lee, M., Borom, M. P. and Szala, L. E., Rapid rate sintering of ceramics, U. S. Patent 4490319, 1984.

32. Kim, D. H. and Kim, C. H., Effect of heating rate on pore shrinkage in yttria-doped zirconia, *J. Am. Ceram. Soc.*, **76**, 1877–78, 1993.

33. Yoo, Y.-S., Kim, J.-J. and Kim, D.-Y., Effect of heating rate on the microstructural evolution during sintering of BaTiO$_3$ ceramics, *J. Am. Ceram. Soc.*, **70**, C322–24, 1987.

34. Huckabee, M. L. and Palmour III, H., Rate controlled sintering of fine grained Al$_2$O$_3$, *Am. Ceram. Soc. Bull.*, **51**, 574–76, 1972.

35. Cannon, R. M., Rhodes, W. H. and Heuer, A. H., Plastic deformation of fine-grained alumina (Al$_2$O$_3$): I, interface controlled diffusional creep, *J. Am. Ceram. Soc.*, **63**, 46–53, 1980.

36. Gupta, T. K., Instability of cylindrical voids in alumina, *J. Am. Ceram. Soc.*, **61**, 191–95, 1978.

37. Rhee, S. K., Critical surface energies of Al_2O_3 and graphite, *J. Am. Ceram. Soc.*, **55**, 300–303, 1972.

38. Kingery, W. D., Metal-ceramic interactions: IV, Absolute measurements of metal-ceramic interfacial energies, *J. Am. Ceram. Soc.*, **37**, 42–45, 1954.

39. Morgan, C. S. and Tennery, V. J., Magnesium oxide enhancement of sintering of alumina, in *Sintering Processes*, Mater. Sci. Res. Vol. 13, G. C. Kuczynski (ed.), Plenum Press, New York, 427–36, 1980.

PART V
SINTERING OF IONIC COMPOUNDS

In the previous Parts, II to IV, we considered only systems where densification and grain growth occur by movement of atoms of a specific (single) component. In addition, the vacancy concentration in a specific region was assumed to be determined only by the capillary pressure of the region arising from its geometry. However, the vacancy concentration in most ceramics, in particular ionic compounds, varies with the addition of dopants (also called additives, in general) and accordingly the sinterability also varies. Furthermore, depending on point defect concentration and temperature, the species that determines sinterability can also change. In Part V, we will consider the densification and grain growth of ionic compounds in relation to defect chemistry, ion diffusion and ion segregation.

PART V
SINTERING OF IONIC COMPOUNDS

12

SINTERING ADDITIVES AND DEFECT CHEMISTRY

Sintering additives are usually added to powders in an attempt to enhance the sinterability and to control the microstructure. Typical examples are the addition of Ni to W for improving sinterability, and of MgO to Al_2O_3 for suppressing abnormal grain growth and improving densification. However, for the most part the roles of sintering additives are only known empirically and their mechanisms are not well understood. This chapter considers the point defects formed by the addition of sintering additives* in ionic compounds with low defect concentrations. For a low concentration of point defects, we may assume that the matrix atoms and point defects form an ideal solution with no interaction between defects. We may assume also that the concentration of matrix atoms is 1. In this case, we can easily estimate the concentration of point defects caused by dopant addition. Therefore, in the case of lattice-diffusion-controlled sintering, the estimation can explain the change in sinterability with dopant addition.

12.1 POINT DEFECTS IN CERAMICS

Point defects in ceramics are usually expressed using the Kröger–Vink notation.[1,2] According to this notation, addition (or depletion) of neutral atoms and free electrons are expressed separately. The atoms and defects are denoted by alphabetic characters, their locations by subscripts and their effective charge by superscripts; namely, in the form of A_B^C where A means a specific atom or defect, B its location and C its effective charge. A positive effective charge is denoted as •, a negative effective charge as $\prime$, and a neutral (zero) charge as x or with nothing. Table 12.1 lists some typical point defects present in a compound MX.[3–5] Among those listed, the two most common

---

*In ionic compounds, sintering additives are usually called dopants when their concentration is low.

Table 12.1. Various types of point defects in a compound MX

Defect type	Symbols
Vacancies	$V_M, V_X, V_M^{\bullet}, V_X^{\bullet}, V_M'', V_X^{\bullet\bullet}, \ldots$
Interstitials	$M_i, X_i, M_i^{\bullet\bullet}, X_i'', \ldots$
Misplaced atoms	$X_M, M_X, \ldots$
Associated centres	$(V_M\, V_X), (X_i\, X_M), \ldots$
Foreign atoms	$L_M, L_i^{\bullet\bullet}, F_M^{\bullet}, \ldots$
Free electrons and holes	$e', h^{\bullet}$

and important types of intrinsic crystalline defects in an ionic compound are Frenkel and Schottky defects. In addition to ionic defects, electronic defects (free electrons and electron holes) are also available in ionic compounds.

12.1.1 Frenkel Defect

A Frenkel defect forms when an atom (ion) goes into an interstitial site leaving behind a vacancy and therefore consists of a defect pair* of an interstitial atom and its vacancy. For a simple metal oxide compound MO with full ionization, its formation equation is expressed as

$$M_M^X \rightleftarrows M_i^{\bullet\bullet} + V_M'' \qquad (12.1)$$

When the number of these defects is very low compared with the number of lattice points, elementary statistical mechanics or the conventional mass action law gives

$$[M_i^{\bullet\bullet}][V_M''] = \exp\left(-\frac{\Delta g_F}{kT}\right) = K_F \qquad (12.2)$$

where $[M_i^{\bullet\bullet}]$ is the concentration of interstitial atoms with an effective charge of $+2$, Δg_F the formation free energy of a Frenkel defect and K_F its mass action constant.

12.1.2 Schottky Defect

The Schottky defect, which is unique to ionic compounds, consists of a stoichiometric pair of cation and anion vacancies. For an MO compound with full ionization, its formation equation is expressed as

$$M_M + O_O \rightleftarrows V_M'' + V_O^{\bullet\bullet} + M_B + O_B \qquad (12.3)$$

---

*Here, 'pair' does not mean an interstitial–vacancy associate but the two separated conjugate defects, an interstitial and its vacancy.

Here, B denotes the place where a lattice can form, for example, the grain boundary, a surface or a dislocation. Hence, unlike the Frenkel defect, the Schottky defect creates new lattice sites. Since M_M and O_O are equivalent to M_B and O_B, respectively, Eq. (12.3) can also be written as

$$null \rightleftarrows V''_M + V^{\bullet\bullet}_O \tag{12.4}$$

The concentrations of these defects are then expressed as

$$[V''_M][V^{\bullet\bullet}_O] = \exp\left(-\frac{\Delta g_S}{kT}\right) = K_S \tag{12.5}$$

where Δg_S is the formation free energy of a Schottky defect and K_S its mass action constant.

12.1.3 Electronic Defect

Perfect electronic order is achieved only at a temperature of 0K, where all electrons are in the lowest possible energy levels under the constraint of the Pauli exclusion principle. Any excitation of electrons from their ground state to higher energy levels results in electronic disorder. In ceramics, however, intrinsic electronic disorder refers to the formation of free electrons in the conduction band and holes in the valence band. An intrinsic electronic defect thus consists of a free electron in the conduction band and a free electron hole in the valence band. The concentrations of free electrons (e') and electron holes ($h^{\bullet}$) are determined by the band gap and temperature. According to Fermi statistics, the probability of an electron occupying an energy level E, $P(E)$, is expressed as

$$P(E) = \frac{1}{1 + \exp[(E - E_F)/kT]} \tag{12.6}$$

where E_F is the Fermi level. In an intrinsic insulator or semiconductor, the concentration of free electrons, n, is the same as that of free electron holes, p. Denoting E_c to be the energy level of the conduction band and E_v that of the valence band, E_F is $\sim(E_v + E_c)/2$. When the concentrations of free electrons and free electron holes are low, Eq. (12.6) gives

$$n = [e'] = \frac{n_e}{N_c} \cong \exp\left[-\frac{(E_c - E_F)}{kT}\right] \tag{12.7}$$

and

$$p = [h^{\bullet}] = \frac{n_h}{N_v} \cong \exp\left[-\frac{(E_F - E_v)}{kT}\right] \tag{12.8}$$

Here, n_e and n_h are, respectively, the number of electrons and holes per unit volume in the conduction and valence band, and N_c and N_v are, respectively, the density of the electron state in the conduction band and that of the electron hole state in the valence band. According to elementary quantum physics, they are expressed as

$$N_c = 2\left(\frac{2\pi m_e^* kT}{h_p^2}\right)^{3/2} \tag{12.9}$$

and

$$N_v = 2\left(\frac{2\pi m_h^* kT}{h_p^2}\right)^{3/2} \tag{12.10}$$

where h_p is the Planck constant $(6.623 \times 10^{-34}\,\text{J s})$ and m_e^* and m_h^* are respectively the effective masses of a free electron and an electron hole.

When external conditions, for example, solute addition and non-stoichiometry, dominate the electronic defects (extrinsic electronic defects), the electronic energy levels change and the relative concentrations of free electrons and electron holes vary. (Non-stoichiometry means a change in stoichiometry caused by a change in external conditions, for example, atmosphere and impurities.) However, the product of the two concentrations, $[e'][h^\bullet]$ at a given temperature is constant, namely,

$$np = [e'][h^\bullet] = \left[\exp\left(-\frac{E_g}{kT}\right)\right] = K_i \tag{12.11}$$

where K_i is the mass action constant of the electronic defect and $E_g \equiv E_c - E_v$.

12.2 FORMATION OF POINT DEFECTS BY ADDITIVES

The concentrations of point defects in ionic compounds vary with the concentration of dopants. The defect concentrations can be predicted using defect chemistry[1-3] where each defect species is considered to be a chemical species and the reactions among defects are expressed as chemical reaction equations. In expressing defect chemical reactions, some basic principles are applied.

First, the mass action law must be satisfied between defect species.

Second, the sum of effective charges on the left-hand side of any reaction equation must be the same as that on the right-hand side. Accordingly, an overall electrical neutrality condition is maintained in the sample.

Third, the cation to anion site ratio must be constant even for non-stoichiometry (site relation). In other words, as long as the crystal structure of an M_aX_b compound is maintained, the ratio a/b is invariable. However, the absolute number of sites may vary according to defect reactions. In respect of the site relation, there are species which create lattice sites, for example, V_M, V_X, M_M, X_M and X_X, and species which do not create lattice sites, for example, e, h, M_i and L_i.

From the above principles, we can predict the concentrations of various defects using

  (i)   the equations governing the formation of ionic defects,
  (ii) the equation governing the formation of electronic defects,
  (iii) a reaction equation between material and atmosphere,
  (iv) a mass conservation equation of the dopant, and
  (v)   an electrical neutrality equation (condition) of the total defects.

The equations for (i)–(iii) are expressed as products of the defect concentrations and the equations for (iv) and (v) as sums. We can calculate the concentration of each defect from the equations exactly by using a personal computer with n equations for n concentration variables. However, the usual and simple method of calculation follows a suggestion of Brouwer.[1] Since the concentrations of defects vary drastically with external (thermodynamic) conditions over a range of several orders of magnitude, the concentrations of minor species can be ignored and only the concentration of the major defect species considered for the equations for (iv) and (v). These are then expressed in the form of concentration products and the variation of defect concentration with external condition is easily shown in a log–log scale, the Brouwer diagram.[1] Among the equations for (i)–(v), those for (i)–(iii), which follow the mass action law, hold irrespective of temperature, T, partial pressure of vapourizable species, p_a, and dopant concentration, C_L. In contrast, when using equations for (iv) and (v), we first assume the defect type of the dopant in the material and the charge neutrality between the major defects. In other words, major defects in the equations for (iv) and (v) vary with T, p_a and C_L; the assumption is correct only under specific conditions. Therefore, if the predicted results are not in accord with observed results, this means that the assumptions made for major defects in the equations for (iv) and (v) were incorrect. This section will consider only the effect of dopant concentration on the concentrations of other defects in systems where T and p_a are constant.

Consider a case where L_2O_3 dopant is added to MO oxide. A reaction of MO with oxygen in the atmosphere occurs following

$$\frac{1}{2}O_2 \rightarrow O_O^x + V_M'' + 2h^\bullet \qquad (12.12)$$

if an oxygen atom goes into the oxide forming a new lattice site. (This is the usual case in oxides.) Then, following the mass action law, Eq. (12.13) holds.

$$\frac{[O_O^x][V_M'']p^2}{p_{O_2}^{1/2}} \approx \frac{[V_M'']p^2}{p_{O_2}^{1/2}} = K_g \tag{12.13}$$

Here, K_g is the reaction constant. Therefore, the equations for (i)–(iii) are Eqs (12.2), (12.5), (12.11) and (12.13).

If all of the dopant atoms L go to M sites in the lattice as donors,

$$L_2O_3 \xrightarrow{MO} 2L_M^\bullet + V_M'' + 3O_O^x \tag{12.14}$$

is satisfied. From this equation, the equation for (iv) becomes

$$[L]_{total} = [L_M^\bullet] \tag{12.15}$$

On the other hand, the electrical neutrality condition gives

$$[L_M^\bullet] + p + 2[V_O^{\bullet\bullet}] + 2[M_i^{\bullet\bullet}] = 2[V_M''] + n \tag{12.16}$$

If the major defects in MO are of the Schottky type,

$$[V_O^{\bullet\bullet}] \approx [V_M''] \tag{12.17}$$

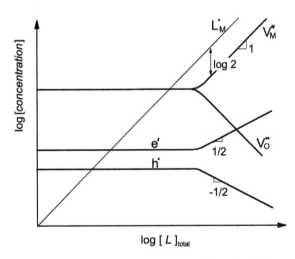

Figure 12.1. Effect of the concentration of foreign atom L on the defect situation in a compound MO; p_{O_2} = const., $[V_O^{\bullet\bullet}] \geq [V_M'']$ and $n \geq p^3$.

for the intrinsic region and

$$[L_M^{\bullet}] \approx 2[V_M''] \tag{12.18}$$

for the extrinsic region. Therefore, taking logarithmic forms of Eqs (12.5), (12.11), (12.13), (12.15), (12.17) and (12.18) gives the relations between dopant concentration and the other defect concentrations, as shown in Figure 12.1.[3] This figure corresponds to a system where $[V_O^{\bullet\bullet}] \gtrsim [V_M'']$ and $n \gtrsim p$ satisfy the case of pure MO under a given P_{O_2}. In Figure 12.1, we observe that addition of a dopant increases the concentration of defects with opposite charges and decreases that of those with similar charges. This result is evident from the reaction equations and also from the Le Chatelier principle. If the lattice diffusion of M controls the sintering of this material, sinterability is expected to increase with the addition of L_2O_3, which increases the metal vacancy concentration $[V_M'']$ and hence the lattice diffusion coefficient D_M.

13

DENSIFICATION AND GRAIN GROWTH IN IONIC COMPOUNDS

The transport of ions during the densification of ionic compounds occurs not only by a chemical potential gradient coming from a difference in capillary pressure but also an electrical potential gradient derived from the difference in diffusivity between different ions. On the other hand, an electrostatic potential in the material affects the segregation of ions at the grain boundary, and further affects the grain boundary migration and grain growth.

13.1 DIFFUSION AND SINTERING IN IONIC COMPOUNDS

During the densification of ionic compounds material transport occurs mostly by diffusion with the material maintaining its stoichiometry, in other words, by diffusion, in effect, of lattice molecules. The diffusion of each ion species occurs under a chemical potential gradient derived from a capillary pressure difference and an electrical potential gradient resulting from a difference in mobility between the cation and the anion. The diffusion fluxes of cations and anions are in fact interrelated and occur in the same direction, i.e. ambipolar diffusion. The driving force for the diffusion of ions is thus the electrochemical potential gradient $\nabla \eta$, the sum of the chemical potential gradient and the electrical potential gradient. The diffusion flux of ionic species i, J_i, is then expressed as

$$J_i = C_i v_i = -C_i B_i \nabla \eta_i$$
$$= -C_i B_i [\nabla \mu_i + Z_i F \nabla \varphi] \tag{13.1}$$

where C_i is the (molar) concentration, B_i the (mechanical) mobility, μ_i the chemical potential, Z_i the effective charge, F the Faraday constant (96486.7 Coulomb/mole), and φ the electrical potential.[6] For an $M_a X_b$ compound with a cation valence of b and anion valence of $-a$

$$J_M = -\frac{C_M D_M}{RT}[\nabla\mu_M + bF\nabla\varphi] \tag{13.2a}$$

$$J_X = -\frac{C_X D_X}{RT}[\nabla\mu_X - aF\nabla\varphi] \tag{13.2b}$$

where $D_i(i = M^{b+}, X^{a-})$ is the self-diffusion coefficient, and M and X denote cation M^{b+} and anion X^{a-}, respectively. When transport of electrons and holes is negligibly slow compared to transport of cations and anions, the ionic fluxes are coupled via an electro-neutrality field or $\sum Z_i J_i = 0$, leading to a flux of lattice molecules $M_a X_b$.

The flux equation of lattice molecules $M_a X_b$ can be derived using the flux and stoichiometry constraints,

$$J_{M_a X_b} = \frac{1}{a}J_M = \frac{1}{b}J_X \tag{13.3}$$

and

$$C_{M_a X_b} = \frac{1}{a}C_M = \frac{1}{b}C_X \tag{13.4}$$

and the general relationship

$$\nabla\mu_{M_a X_b} = a\nabla\mu_M + b\nabla\mu_X \tag{13.5}$$

from the consideration of local thermodynamic equilibrium. Combining Eqs (13.3)–(13.5) with Eq. (13.2) gives

$$J_{M_a X_b} = -\frac{C_{M_a X_b}}{RT}\left(\frac{D_M D_X}{bD_M + aD_X}\right)\nabla\mu_{M_a X_b}$$
$$\equiv -\frac{C_{M_a X_b}}{RT}\overline{D}\nabla\mu_{M_a X_b} \tag{13.6}$$

where $\overline{D}$ is the effective self-diffusion coefficient of the lattice molecule. If the thermodynamic factor of $M_a X_b$ ($\equiv \partial\mu_{M_a X_b}/\partial C_{M_a X_b}$) equals unity (this assumption would be reasonable, as in an elemental metal with a gradient of vacancy concentration), Eq. (13.6) can be expressed in the form of Fick's first law as

$$J_{M_a X_b} = -\left(\frac{D_M D_X}{bD_M + aD_X}\right)\nabla C_{M_a X_b}$$
$$\equiv -\overline{D}\nabla C_{M_a X_b} \tag{13.7}$$

The flux equations for cations and anions in the form of Fick's first law can also be obtained from Eq. (13.7) using Eqs (13.3) and (13.4); they take a similar form to Eq. (13.7) with the identical effective diffusivity $\overline{D}$.

Concerning the effective diffusion coefficient, two types of expression have been, reported so far.[7-11] One is[7-9]

$$\overline{D} = \frac{D_M D_X}{bD_M + aD_X} \tag{13.8}$$

as derived above, and the other is[10,11]

$$\overline{D} = \frac{(a+b)D_M D_X}{bD_M + aD_X} \tag{13.9}$$

For the derivation of Eq. (13.9) use is made of an expression $\nabla \mu_i = RT \times (\nabla C_i / C_i)$ for an ionic species i, which assumes an ideal solution between M^{b+} and X^{a-}, in addition to Eqs (13.3) and (13.4). This assumption, however, cannot be justified because 'ionic solution' deviates significantly in a negative fashion from the ideal solution behaviour. This means that the effective diffusion coefficient in an M_aX_b compound should be expressed as Eq. (13.8) and the flux equation as Eq. (13.6) in general or as Eq. (13.7). Equation (13.8) shows that if there is a large difference in diffusion coefficient between the two different ions, the slower moving ion governs the effective diffusion coefficient and the densification during sintering. Figure 13.1 plots the temperature dependence of $\overline{D}$ on a log D versus $1/T$ plane.

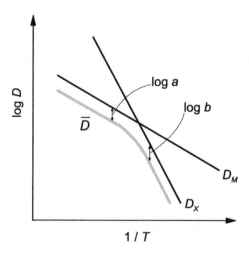

Figure 13.1. Effective diffusion coefficient $\overline{D}$ of an M_aX_b compound where the activation energy of the cation diffusion is lower than that of the anion diffusion.

When a compound is sintered via diffusion through various paths, the effective diffusivity is the sum of each contribution through its specific path.[6,9] The effective diffusion coefficient of ion i, $\overline{D}_i$, is then expressed as

$$\overline{D}_i = \sum_p (D_i f)^p \qquad (13.10)$$

where f is the area fraction of each path p. When both lattice and grain boundary diffusion occur simultaneously and contribute to effective diffusivity,*

$$\overline{D}_i = (D_i f)^l + (D_i f)^b \qquad (13.11)$$

where superscript l denotes the lattice and b the boundary. From Eq. (13.3), $\overline{D}$ is expressed as

$$\overline{D} = \frac{(D_M^b f^b + D_M^l f^l)(D_X^b f^b + D_X^l f^l)}{b(D_M^b f^b + D_M^l f^l) + a(D_X^b f^b + D_X^l f^l)} \qquad (13.13)$$

The contribution of grain boundary diffusion increases as the grain boundary area increases, i.e. the grain size decreases following $f^b \propto 1/G$ (Eq. (13.12)). Unlike grain boundary diffusion, the contribution of lattice diffusion can be considered to be independent of grain size if the volume fraction of the grain boundary is inconsiderable. For an MX compound, if $D_X^b > D_M^b$ and $D_X^l < D_M^l$, $\overline{D}$ varies with grain size, as shown in Figure 13.2. Sintering is thus governed by the slowest moving species over its fastest path.[13,14]

In real sintering, however, the material flux coming from grain boundaries to a pore is affected also by the surface area of the pore for lattice diffusion and the grain boundary length at the pore surface for grain boundary diffusion (see Section 4.2 and Section 5.2). Therefore, the densification of an ionic compound may not be governed by the fastest path of the slowest moving species. In particular, the final densification of small pores was recently predicted to be governed always by grain boundary diffusion.[15]

In the case of lattice diffusion, the effective diffusion coefficient can vary considerably with the addition of a dopant because the dopant addition can drastically change the concentration of defects, as explained in Chapter 12. Consider, as an example, an MO oxide with vacancies as major defects. If the activation energy of cation vacancy diffusion, Q_{V_M}, is lower than that of anion vacancy diffusion, Q_{V_X}, the maximum effective diffusivity appears in the region with oxygen deficiency (Figure 13.3).[16] When interstitial cations also contribute

---

*The effective diffusion coefficient in a polycrystal of grains with a size G may be expressed as[12]

$$\overline{D} = D_l + \frac{\pi \delta_b D_b}{G} \qquad (13.12)$$

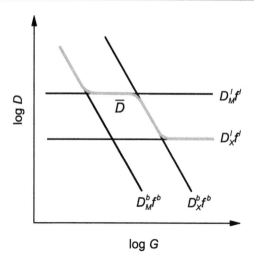

Figure 13.2. Relationship for ambipolar diffusion in an MX compound: effective diffusion coefficient versus grain size when both lattice and grain boundary diffusion are operative.

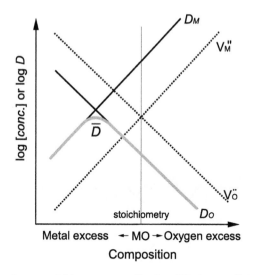

Figure 13.3. Effect of non-stoichiometry on effective diffusion coefficient in an MO oxide where the Schottky defect is dominant.

to ion diffusion, the situation becomes complicated. Figure 13.4 shows an example where oxygen ion diffusion through oxygen vacancies controls the effective diffusivity of sintering when the activation energy of cation interstitial diffusion, Q_{M_i}, is the lowest ($Q_{M_i} < Q_{V_M} < Q_{V_X}$).

In discussing the effect of dopants on densification using defect chemistry, several points need to be considered.

(1) Since the dopant increases the concentration of defects with opposite charges and decreases the concentration of those with like charges, the effects on the concentrations of ion vacancies and interstitial ions are always opposite. Therefore, if $Q_{M_i} \ll Q_{V_M}$ in an ionic compound with Schottky defects as major defects and Frenkel defects as minor defects, the effective diffusion coefficient of the cation can vary from D_{M_i} to D_{V_M} as the concentration of donor dopant increases.[17] This result can also be seen in Figure 13.4 where the species that determines D_M changes from the cation interstitial to the cation vacancy as the concentration of cation vacancies increases.

(2) The effect of impurities cannot be ignored. In reality, not only dopants but also impurities can change defect concentrations and the effective diffusivity considerably. Sometimes impurities dominate the overall phenomenon. In particular, this may occur in systems with a very low dopant solubility.

(3) Several equations of defect equilibria are possible for the same dopant and sometimes different equations can predict the same effect of dopant on effective diffusivity. In such a case, additional experiments are needed to identify the appropriate defect equilibria.

(4) The use of defect chemistry for predicting effective diffusivity and sintering kinetics has limitations because of the complex effects of dopant and impurities. The dopant can cause secondary effects, including changes in grain boundary diffusivity, grain boundary energy, surface energy and grain boundary mobility. In particular, a dopant can change the

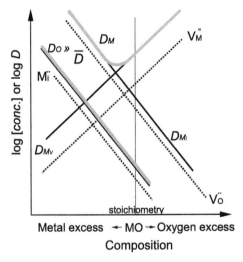

Figure 13.4. Effect of non-stoichiometry on effective diffusion coefficient in an MO oxide where both the Schottky and Frenkel defects are considerable.

anisotropy in grain boundary and surface energy considerably as recently shown in Al_2O_3[18,19] and $SrTiO_3$.[20,21] With a change in anisotropy, grain growth and densification behaviour vary remarkably (see Section 9.2.1 and Section 10.1). It follows that the effect of a dopant is much more complicated than anticipated by defect chemistry and is mostly evaluated empirically.

For a system with a known ion species and the defect that governs the effective diffusivity, the selection of a dopant follows, in general, the following rules:[22]

    (i) the valence rule: the effective valence of a dopant ion is different by one (or two) from that of the host ion;

   (ii) the size rule: the size of a dopant ion is similar to that of the host ion in order to replace it;

 (iii) the concentration rule: the amount of dopant is usually within the limit of its solubility.

13.2 ELECTROSTATIC POTENTIAL EFFECT ON INTERFACE SEGREGATION

13.2.1 Pure Material without Dopant

In pure ionic compounds, thermally generated defects are present maintaining an electrical neutrality condition. Consider a univalent MX compound with Schottky disorder as the major defects. The formation free energy of a cation vacancy, g_{V_M}, is, in general, lower than that of an anion vacancy, g_{V_X}. However, since the electrical neutrality condition is maintained in the lattice, the number of cation and anion vacancies is the same, which is possible in the presence of an electrostatic potential φ in the bulk. On the other hand, in the simplest formulation, the surface and the grain boundary (more generally, interface) can be regarded as an infinite and uniform source and sink of vacancies (see Section 4.1.4). This means that the interface provides ions with an infinite number of sites of the same energy level when they come to the interface forming vacancies at their original lattice sites following

$$M_M \rightleftharpoons M_b^\bullet + V_M' \tag{13.14}$$

$$X_X \rightleftharpoons X_b' + V_X^\bullet \tag{13.15}$$

where $M_b^\bullet$ denotes ion M at the boundary (interface). Therefore, the electrical neutrality condition does not apply to the interface and the concentrations of cation and anion vacancies are determined by their formation free energy at the interface. In other words, φ is zero at the interface and varies with distance d from the interface $(\varphi(d))$.[23]

The concentrations of cation and anion vacancies are then expressed, respectively, as

$$[V'_M] = \exp\left(-\frac{g_{V_M} - Ze\varphi(d)}{kT}\right) \tag{13.16}$$

$$[V^{\bullet}_X] = \exp\left(-\frac{g_{V_X} + Ze\varphi(d)}{kT}\right) \tag{13.17}$$

where $Z (=1$ in this case) is the effective charge of the defect. At $d=0$,

$$[V'_M] = \exp\left(-\frac{g_{V_M}}{kT}\right) \tag{13.18}$$

$$[V^{\bullet}_X] = \exp\left(-\frac{g_{V_X}}{kT}\right) \tag{13.19}$$

However, at $d = \infty$,

$$[V'_M]_\infty = [V^{\bullet}_X]_\infty = \exp\left(-\frac{1}{2}\frac{(g_{V_M} + g_{V_X})}{kT}\right) \tag{13.20}$$

and

$$e\varphi_\infty = \frac{1}{2}(g_{V_M} - g_{V_X}) \tag{13.21}$$

For NaCl, as an example, an estimate taking $g_{V_{Na}} = 0.65eV$ and $g_{V_{Cl}} = 1.21eV$ gives $\varphi_\infty = -0.28V$.[24] Since the concentrations of cation and anion vacancies formed at and near the interface are different from each other, ionic vacancies by themselves cannot satisfy the electrical neutrality. However, this electrical imbalance is compensated for by the segregation of ions of one kind at the interface. Figure 13.5 illustrates schematically the segregation of cations at an interface and the concentration profiles of cation and anion vacancies in an ionic compound with $g_{V_M} < g_{V_X}$. The near-interface region with a difference in vacancy concentration is referred to as the space charge region. Considering the excess cation vacancies in this region and the cations segregated at the interface, the electrical neutrality condition is satisfied in the interface region which includes the interface and the space charge region. In this regard the near-interface deviations from stoichiometry in a pure ionic compound do not constitute an interface segregation phenomenon in a Gibbsian concept of the dividing interface (see Section 2.1).

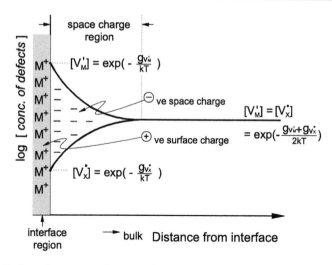

Figure 13.5. Boundary space charge and associated charged defect distributions for an MX compound with $g_{V_M'} < g_{V_X^\bullet}$.

13.2.2 Impure Material with Dopant

Aliovalent dopants or impurities in ionic compounds change the space charge level and interact with interfaces. Doping an aliovalent solute with a positive effective charge, for example $CaCl_2$ to $NaCl$ or Al_2O_3 to MgO, results in the formation of extrinsic cation vacancies in addition to intrinsic cation vacancies. Equation (13.16) holds even for the case where the concentration of the vacancies generated by the addition of aliovalent solute L is higher than the intrinsic concentration determined by thermal energy. For a univalent MX compound with a solute of single effective charge, the cation vacancy concentration is expressed as

$$C_L \approx [V_M']_\infty = \exp\left(-\frac{g_{V_M} - e\varphi_\infty}{kT}\right) \tag{13.22}$$

where C_L is the concentration of the aliovalent solute. Here, the sign and the absolute value of the electrostatic potential φ_∞ is dependent on temperature and solute (dopant) concentration.

When $CaCl_2$ is added to $NaCl$, as an example, Ca ions go into Na sites following the equation

$$CaCl_2 \xrightarrow{2NaCl} Ca_{Na}^\bullet + V_{Na}' + 2Cl_{Cl}^X \tag{13.23}$$

Then, with the addition of $CaCl_2$, $[V_{Na}']$ and $[Cl_b']$ increase and $[V_{Cl}^\bullet]$ and $[Na_b^\bullet]$ decrease according to the equation for Schottky defect formation (Eq. (12.4)) and the equations for cation and anion vacancy formation (Eqs (13.14) and

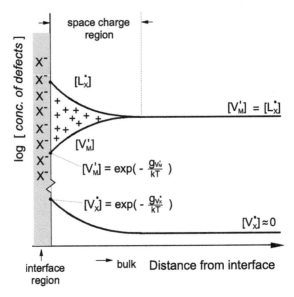

Figure 13.6. Boundary space charge and associated charged defect distributions for an MX compound containing aliovalent solute L with a positive effective charge.

(13.15)). For a high $[Ca_{Na}^{\bullet}]$, φ_{∞} is positive and the boundary potential negative. For ionic compounds, φ_{∞} is typically in the range between ± 0.1 and ± 1.0 V.[25]

Figure 13.6 schematically shows the distributions of defects with distance from an interface in an MX ionic compound with cation dopant L for an extrinsic region. The dopant addition makes the space charge region positive and the interface negative with anion segregation. Assuming the interface charge is limited to the interface, the depth of the space charge region is characterized by the Debye length l_D[23,25]:

$$ l_D \approx \left(\frac{\varepsilon kT}{8\pi q^2 C_{\infty}} \right)^{1/2} \tag{13.24} $$

where ε is the dielectric constant of the material, q the charge of the defect and C_{∞} the bulk concentration of the defect with a charge of q. For oxide materials with a low concentration of dopants at their sintering temperatures, typical l_D values are in the range between 1 and 10 nm.[23,25,26]

In discussing the defect formation and distribution so far, we assumed, for simplicity, that the interface acts as a perfect source and sink of vacancies and atoms, and provides an infinite number of sites for ions with a charge opposite to the excess charge in the space charge region. However, this assumption is too simple to quantitatively describe the real phenomena. In reality, the number of sites for ions at the interface is limited and the ions at the interface interact with each other. Because of this site limitation, the electrostatic potential in the bulk decreases considerably as temperature increases.[27]

13.3 SOLUTE SEGREGATION AND GRAIN BOUNDARY MOBILITY

The amount of grain boundary segregation in ionic compounds can also be estimated using the regular solution formalism,[25,28,29] as in Section 7.1. In their bond model approach, Wynblatt and McCune[28,29] suggested that, in addition to an interfacial energy contribution, mixing enthalpy contribution and elastic strain energy contribution, an electrostatic contribution should be included in the heat of segregation for the case of aliovalent solutes. Table 13.1 lists the estimated ranges of the four kinds of contributions to the segregation enthalpy in metal oxides.[28] Among the four contributions, the effects of strain energy and electrostatic interaction energy are usually predominant and interact with each other.[28-30]

In real systems, however, the segregation behaviour is much more complicated than the simple understanding based on the bond model approach of a single segregant. With multiple segregants and impurities, interactions among them would result in complicated segregation behaviour. In the case of Nb-doped $SrTiO_3$, Nb segregation at grain boundaries was observed in H_2 but not observed in air.[31,32] The Nb segregation in H_2 was attributed to the formation of a negative grain boundary core due to the segregation of inherently present acceptor impurities and electrons trapped at grain boundaries. This example shows that the grain boundary segregation in oxides can vary considerably not only with impurities but also with oxygen partial pressure.

Experimental evidences of specific ion or solute segregation and its effect on grain boundary mobility are abundant, as introduced in Section 7.2. With the segregation of solute ions, the boundary velocity can decrease considerably[32] as a result of solute drag within a low velocity limit described in Chapter 7. Segregation of a specific ion species with an electric charge was further confirmed in a recent investigation on the effect of an electric field on

Table 13.1. Approximate ranges of contribution to the heat of segregation of solutes in metal oxides (kJ/mol)[28]

Interfacial energy (ΔH_γ)	Binary interaction (ΔH_m)	Strain energy (ΔH_ε)	Electrostatic interaction (ΔH_e)
$0 \sim \pm 20^{(a)}$	$0 \sim \pm 60^{(b)}$	$0 \sim -140^{(c)}$	$0 \sim \pm 100^{(d)}$

(a) Assuming a maximum interface energy difference of $0.5 \, J/m^2$.
(b) Based on heats of formation of various spinels.
(c) Computed from Eq. (7.4) with reasonable limiting values of the parameters.
(d) For $\varphi_\infty \approx 0.5V$ (corresponding to low temperatures) and an electronic charge difference of 2.

boundary mobility.[33] In a bi-layer of large and small Al_2O_3 grains, the growth of large grains towards the layer with small grains was either accelerated or retarded, respectively, when applying a positive or a negative bias to the small-grained layer. This result shows that the mobility of grain boundaries with segregation of charged ions can be affected considerably by an external electric field.

PROBLEMS

5.1. Using a statistical treatment, show that the vacancy concentration, n_v/N_A, in a pure compound with Frenkel defects is expressed as

$$\frac{n_v}{N_A} = \exp\left(-\frac{\Delta g_F}{2kT}\right)$$

where n_v is the number of vacancies per mole, N_A the Avogadro number and Δg_F the formation free energy of a Frenkel defect.

5.2. When the oxygen concentration of an air-sintered pure oxide is decreased by annealing in a reducing atmosphere, by how much is the Fermi level of this oxide changed?

5.3. Consider the addition of an acceptor dopant MO to L_2O_3 oxide. Show the change in defect concentrations with dopant amount. Assume that the major and minor defects in L_2O_3 are the Schottky defect and the electronic defect, respectively, and that all M cations go into L sites.

5.4. Show the change in defect concentrations in Al_2O_3 with an increased addition of ZrO_2. Assume that the major and minor defects in Al_2O_3 are Schottky and Frenkel defects, respectively, and that all Zr cations go into Al sites.

5.5. In ZrO_2-doped Al_2O_3 (refer to Problem 5.4), how do you expect the effective diffusion coefficient of Al to vary by increasing the amount of ZrO_2 under the conditions where the activation energy of Al vacancy diffusion is much smaller than that of Al interstitial diffusion, and vice versa?

5.6. L_2O_3 dopant is added to an MO oxide with Schottky defects as the major defects. Assuming that all the L ions go into M sites,
   (a) plot the variation of $[V_M'']$, $[V_O^{\bullet\bullet}]$ and $[L_M^{\bullet}]$ with temperature on a log[concentration] versus $1/T$ plane, and
   (b) plot the effective diffusion coefficient $\bar{D}$ in MO with temperature. Assume that $\bar{D}$ is governed by the diffusion of V_M''.

5.7. The major point defects in NaCl are Schottky defects. When $CaCl_2$ is added to NaCl, Ca ions replace Na ions. For a $CaCl_2$-doped NaCl, plot and explain the variation in $[V_{Na}']$ with temperature in a log[V_{Na}'] versus $1/T$ plane.

5.8. Derive Eq. (13.6).

5.9. Sketch the variation of effective diffusion coefficient with grain size for an M_aX_b compound where both lattice and grain boundary diffusions are operative. Assume $D_M^l > D_X^l$ and $D_M^b > D_X^b$.

5.10. Consider an MO oxide where ions move by the vacancy mechanism and the activation energy of metal ion diffusion, Q_M, is higher than that of oxygen ion diffusion, Q_O, $(Q_M > Q_O)$.

    (a) Sketch the variations of $[V_O^{\bullet\bullet}]$ and $[V_M'']$, and also of the diffusion coefficients of the two different ions with oxygen non-stoichiometry.

    (b) In the sintering of this compound, what are the diffusion coefficient D and molar volume V_m that govern the sintering kinetics in the equation $t \propto 1/(JAV_m)$?

5.11. In an MO metal oxide, the dominant defects are known to be Schottky-type. Assuming that $K_S = 10^{-10}$ and $D_M = 100\,D_O$, what kind of dopant and how much of it must be added to the oxide to get the maximum rate of densification in sintering if the densification occurs by lattice diffusion?

5.12. The major defects in KCl are known to be Schottky-type and the formation free energy of the cation vacancy is lower than that of the anion vacancy. Sketch the surface charge and surface defect concentration in pure KCl and in highly $CaCl_2$-doped KCl. Sketch and explain the change in defect concentrations with temperature (in a scale of $1/T$) in the bulk of a $CaCl_2$-doped KCl.

5.13. For an MX compound with $D_M^l > D_X^l$ and $D_M^b > D_X^b$, sketch and explain the variation in the effective diffusion coefficient with grain size.

5.14. Express the mobility of an oxide grain boundary in which the migration is governed by the diffusion of segregated aliovalent dopant atoms.

REFERENCES

1. Kröger, F. A. and Vink, H. J., Relations between the concentrations of imperfections in crystalline solids, in *Solid State Physics Vol. 3*, F. Seitz and D. Turnbull (eds), Academic Press, New York, 307–435, 1956.
2. Kröger, F. A., *The Chemistry of Imperfect Crystals (2nd revised edition) Vol. 2. Imperfaction Chemistry of Crystalline Solids*, North-Holland Publ., Amsterdam, 1974.
3. Brook, R. J., Defect structure of ceramic materials, Chapter 3 in *Electrical Conductivity in Ceramics and Glass, Part A*, N. M. Tallan (ed.), Marcel Dekker, New York, 179–267, 1983.
4. Kröger, F. A., *The Chemistry of Imperfect Crystals (2nd revised edition) Vol. 2. Imperfaction Chemistry of Crystalline Solids*, North-Holland Publ., Amsterdam, 14, 1974.
5. Kingery, W. D., Bowen, H. K. and Uhlmann, D. R., *Introduction to Ceramics* (2nd edition), John Wiley & Sons, New York, 12–76, 1976.
6. Kingery, W. D., Bowen, H. K. and Uhlmann, D. R., *Introduction to Ceramics* (2nd edition), John Wiley & Sons, New York, 381–447, 1976.
7. Readey, D. W., Chemical potentials and initial sintering in pure metals and ionic compounds, *J. Appl. Phys.*, **39**, 2309–12, 1966.
8. Blakely, J. M. and Li, C.-Y., Changes in morphology of ionic crystals due to capillarity, *Acta Metall.*, **14**, 279–84, 1966.
9. Gordon, R. S., Mass transport in the diffusional creep of ionic solids, *J. Am. Ceram. Soc.*, **56**, 147–52, 1973.
10. Ruoff, A. L., Mass transfer problems in ionic crystals with charge neutrality, *J. Appl. Phys.*, **36**, 2903–905, 1965.
11. Rahaman, M. N., *Ceramic Processing and Sintering* (2nd edition), Marcel Dekker, New York, 462–66, 2003.
12. Raj, R. and Ashby, M. F., On grain boundary sliding and diffusional creep, *Metall. Trans. A*, **2A**, 1113–27, 1971.
13. Cannon, R. M. and Coble, R. L., Paradigms for ceramic powder processing, in *Processing of Crystalline Ceramics*, Mat. Sci. Res. Vol. 11, H. Palmour III, R. F. Davis and T. M. Hare (eds), Plenum Press, New York, 151–70, 1978.
14. Gordon, R. S., Understanding defect structure and mass transport in polycrystalline Al$_2$O$_3$ and MgO via the study of diffusional creep, in *Structure and Properties of MgO Al$_2$O$_3$ Ceramics*, W. D. Kingery (ed.), Am. Ceram. Soc. Inc., 418–37, 1984.
15. Kang, S.-J. L. and Jung, Y.-I., Sintering kinetics at final stage sintering: model calculation and map construction, *Acta Mater.*, **52**, 4573–78, 2004.

16. Reijnen, P. J. L., Non-stoichiometry and sintering in ionic solids, in *Problems of Non-stoichiometry*, A. Rabenau (ed.), North-Holland Publ., Amsterdam, 219–38, 1970.

17. Shewmon, P. G., *Diffusion in Solids* (2nd edition), TMS, Warrendale, PA, 162–64, 1989.

18. Park, C. W. and Yoon, D. Y., The effect of SiO_2, CaO and MgO additions on the grain growth of alumina, *J. Am. Ceram. Soc.*, **83**, 2605–609, 2000.

19. Park, C. W. and Yoon, D. Y., Abnormal grain growth in alumina with anorthite liquid and the effect of MgO addition, *J. Am. Ceram. Soc.*, **85**, 1585–93, 2002.

20. Chung, S.-Y., Yoon, D. Y. and Kang, S.-J. L., Effects of donor concentration and oxygen partial pressure on interface morphology and grain growth behavior in $SrTiO_3$, *Acta Mater.*, **50**, 3361–71, 2002.

21. Chung, S.-Y. and Kang, S.-J. L., Intergranular amorphous films and dislocation-promoted grain growth, *Acta Mater.*, **51**, 2345–54, 2003.

22. Brook, R. J., Fabrication principles for the production of ceramics with superior mechanical properties, *Proc. Brit. Ceram. Soc.*, **32**, 7–24, 1982.

23. Kliewer, K. L. and Koehler, J. S., Space charge in ionic crystals. I. General approach with application to NaCl, *Phys. Review*, 140, **4A**, A1226–40, 1965.

24. Kingery, W. D., Bowen, H. K. and Uhlmann, D. R., *Introduction to Ceramics* (2nd edition), John Wiley & Sons, New York, 177–216, 1976.

25. Burggraaf, A. J. and Winnubst, A. J. A., Segregation in oxide surfaces; solid electrolytes and mixed conductions, in *Surface and Near-Surface Chemistry of Oxide Materials* (Mater. Sci. Mono. 47), J. Nowotny and L.-C. Dufour (eds), Elsevier Science Publ., Amsterdam, 449–78, 1988.

26. Nowotny, J., Surface and grain boundary segregation in metal oxides, in *Surfaces and Interfaces of Ceramic Materials*, L.-C. Dufour, C. Monty and G. Petot-Ervas (eds), Kluwer Academic Publ., Dordrecht, 205–39, 1989.

27. Poeppel, R. B. and Blakely, J. M., Origin of equilibrium space charge potentials in ionic crystals, *Surface Sci.*, **15**, 507–23, 1969.

28. Wynblatt, P. and McCune, R. C., Chemical aspects of equilibrium segregation to ceramic interfaces, in *Surfaces and Interfaces in Ceramic and Ceramic–Metal Systems*, Mater. Sci. Res. Vol. 14, J. Pask and A. G. Evans (eds), Plenum Press, New York, 83–95, 1981.

29. Wynblatt, P. and McCune, R. C., Surface segregation in metal oxides, in *Surface and Near-Surface Chemistry of Oxide Materials* (Mater. Sci. Mono. 47), J. Nowotny and L.-C. Dufour (eds), Elsevier Science Publ., Amsterdam, 247–79, 1988.

30. Yan, M. F., Cannon, R. M. and Bowen, K. H., Solute segregation at ceramic interfaces, in *Character of Grain Boundaries*, M. F. Yan and A. H. Heuer (eds), Am. Ceram. Soc. Inc., Columbus, OH, 255–73, 1983.

31. Chiang, Y.-M. and Takaki, T., Grain-boundary chemistry of barium titanate and strontium titanate: I. High-temperature equilibrium space charge, *J. Am. Ceram. Soc.*, **73**, 3278–85, 1990.

32. Chung, S.-Y., Kang, S.-J. L. and Dravid, V. P., Effect of sintering atmosphere on grain boundary segregation and grain growth in niobium-doped $SrTiO_3$, *J. Am. Ceram. Soc.*, **85**, 2805–10, 2002.

33. Jeong, J.-W., Han, J.-H. and Kim, D. Y., Effect of electric field on the migration of grain boundaries in alumina, *J. Am. Ceram. Soc.*, **83**, 915–18, 2000.

PART VI
LIQUID PHASE SINTERING

When a powder compact is sintered in the presence of a liquid phase, i.e. in liquid phase sintering, the density of the compact increases and, at the same time, grains grow as in the case of solid state sintering. The phenomenon in which the average grain size increases via growth of large grains and dissolution of small grains in a matrix is referred to as Ostwald ripening. For the simplest case, Ostwald ripening has been theoretically and rigorously analysed (Lifshitz–Slyozov–Wagner (LSW) theory). The grain growth in real systems, however, is sometimes very different from the simple case, often exhibiting abnormal grain growth. Chapter 15 describes theories of grain growth in a matrix; particular emphasis is given to recent explanations of abnormal grain growth. With respect to densification during liquid phase sintering, two theories have been proposed: the contact flattening theory and the pore filling theory. The former is basically a two-particle model, similar to that of solid state sintering. On the other hand, the latter considers both the densification and grain growth, reflecting the real phenomena that occur during liquid phase sintering. In Chapter 16 the two models and theories are critically examined. Microstructure development during liquid phase sintering is then described.

14

BASIS OF LIQUID PHASE SINTERING

14.1 BASIC PHENOMENA OF LIQUID PHASE SINTERING

Liquid phase sintering is a consolidation technique of powder compacts containing more than one component at a temperature above the solidus of the components and hence in the presence of a liquid. Unlike solid state sintering, the microstructure change during liquid phase sintering is fast because of fast material transport through the liquid. The typical densification curve of liquid phase sintering with sintering time is similar to that of solid state sintering (Figure 4.1), as in those of W-Ni-Fe powder compacts in Figure 14.1.[1] However, as shown in Figure 14.1, a considerable densification usually occurs in the solid state during heating to the liquid phase sintering temperature; the initial microstructure of liquid phase sintering is strongly affected by the solid state sintering stage.[1] In particular, when activated sintering occurs as observed in some systems like W-Ni, more than 90% relative density can be achieved during conventional heating to the liquid phase sintering temperature, as the densification curve of a W(1 μm)-Ni-Fe compact shows in Figure 14.1.

As a liquid phase forms during heating of a powder mixture compact, liquid flows into fine capillaries due to the capillary pressure difference between the fine and coarse channels between solid particles. The solid particles can be redistributed by this liquid flow[2–4] and in liquid phase sintering models this phenomenon is referred to as 'particle rearrangement'. The possibility of particle rearrangement by liquid flow relies on various factors including not only liquid volume fraction but also dihedral angle, extent of sintering at the moment of liquid formation and particle size.[5] When the dihedral angle is larger than 0°, however, particle rearrangement is expected to be unlikely. (Most of the sites of the particles that have melted become pores because of the liquid flow into fine capillaries between solid particles. The elimination of these pores governs the overall densification kinetics.) After liquid formation, the compact consists of three phases: solid, liquid and vapour. As sintering

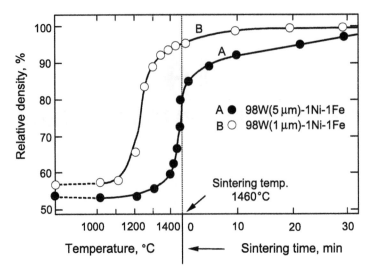

Figure 14.1. Densification curves of 98W-1Ni-1Fe(wt%) alloys during heating to and iso-thermal liquid phase sintering at 1460°C.[1]

proceeds, elimination of pores and growth of grains occur simultaneously in a liquid matrix.

Two models and theories have been proposed to explain densification during liquid phase sintering. The former model, which was proposed by Cannon and Lenel,[6] suggests that liquid phase sintering consists of three stages:

(i) liquid flow,
(ii) solution/reprecipitation, and
(iii) solid state sintering.

Based on this model, Kingery[7] developed a densification theory of liquid phase sintering (see Section 14.2), in particular, for the second stage, by assuming a continuous change in grain shape by a flattening of the contact area between grains, so-called 'contact flattening'. Recent investigations on densification mechanisms,[5,8] however, suggest that contact flattening, the key densification mechanism in Kingery's theory and in its recent modifications,[9,10] is insignificant for densification.

The latter model for densification of liquid phase sintering was proposed by Kwon and Yoon.[11] Based on microstructural observations during liquid phase sintering, they proposed that the filling of pores by the liquid was the essential process for densification and governed the overall sintering kinetics. Later, Park et al.[12] theoretically analysed the liquid filling of isolated pores, and, recently, Kang et al.[8,13] developed a new model and theory (the pore filling theory) of liquid phase sintering. Unlike the previous model[6] and theory,[7] which describe the densification of liquid phase sintering in view of the behaviour of solid grains without taking the grain growth into account, the new model and theory describe the densification in terms of liquid flow into pores as a result of grain growth. (See Section 16.1 and Section 16.2.)

A series of microstructural observations[1,3,14-16] and theoretical analyses[5,8] suggest that densification occurs essentially by the liquid filling of pores. Particle rearrangement due to liquid flow (redistribution) at the very early stage of liquid phase sintering may also contribute to densification under limited and specific conditions.[3-5] Even in this case, however, the contribution would be marginal.[4] During densification by pore filling, grains grow and shape changes occur during growth (see Section 15.2). Therefore, the microstructure development during liquid phase sintering should be analysed with concurrent consideration of densification and grain growth. Basically, normal grain growth in a matrix, either solid or liquid, called 'Ostwald ripening', is explained by the Lifshitz–Slyozov–Wagner (LSW) theory.[17,18]

14.2 CAPILLARITY IN LIQUID PHASE SINTERING

Unlike solid state sintering, liquid flow can occur in liquid phase sintering as a result of capillary action leading to a massive flow of material. Figure 14.2 depicts a small amount of liquid present between two spherical particles. The pressure in the liquid is affected by the geometry of the system, which includes the liquid volume fraction f_l, particle radius a, interparticle distance l, and wetting angle θ. The compressive pressure F between the two particles due to the presence of the liquid is expressed as the sum of the forces coming from the pressure difference between the liquid and the external atmosphere, F_1, and from the surface tension of the liquid, F_2,[19]

$$
\begin{aligned}
F &= F_1 + F_2 \\
&= \gamma_l \left(\frac{1}{r} - \frac{1}{x} \right) \pi a^2 \sin^2 \psi + \gamma_l 2\pi a \sin \psi \sin(\psi + \theta) \\
&= \gamma_l \left[\pi a^2 \sin^2 \psi \left(\frac{1}{r} - \frac{1}{x} \right) + 2\pi a \sin \psi \sin(\psi + \theta) \right]
\end{aligned}
\tag{14.1}
$$

If the shape of the liquid surface is a fraction of a circle (circle approximation),[19,20]

$$
x = a \sin \psi - \left[a(1 - \cos \psi) + \frac{l}{2} \right] \frac{1 - \sin(\psi + \theta)}{\cos(\psi + \theta)}
\tag{14.2}
$$

and

$$
r = \frac{a(1 - \cos \psi) + l/2}{\cos(\psi + \theta)}
\tag{14.3}
$$

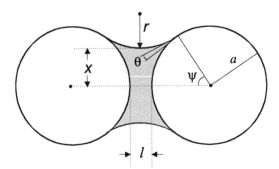

Figure 14.2. Schematic of the contact between two spherical particles with an intermediate liquid film.

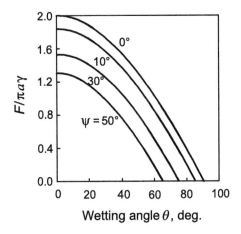

Figure 14.3. Variation of attractive force with the angle of contact for various wetting angles.[19]

Figure 14.3 delineates the calculated variation in compressive pressure between two particles with contact angle ψ and wetting angle θ for interparticle distance $l = 0$.[19] It shows that the compressive pressure increases with decreasing ψ and θ. As the liquid volume fraction approaches zero ($\psi \to 0$), F becomes $2\pi a \gamma_l \cos \theta$. For the case of $l = 0$, the critical wetting angle θ_{cr} for $F = 0$

$$\theta_{cr} = 90 - \frac{\psi}{2} \tag{14.4}$$

For $\theta > \theta_{cr}$, a repulsive force is exerted between the two particles and l becomes larger than zero. For a more general case where $l \neq 0$, the interparticle force can also be calculated as a function of ψ and θ using Eq. (14.1). A calculated result in Figure 14.4 for $\psi = 30°$ shows that the equilibrium interparticle

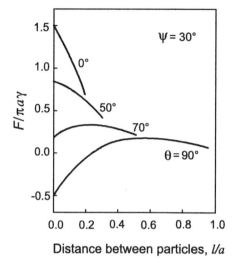

Figure 14.4. Variation of attractive force with particle gap *l* for various wetting angles.[19] Contact angle $\psi = 30°$.

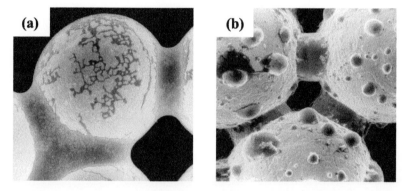

Figure 14.5. Distribution of liquid Cu between W spheres ($\sim$200 μm diameter) for a wetting angle θ of (a) 8° and (b) 85°.[2]

distance increases as the wetting angle increases.[19] For an example of $\theta = 70°$, the equilibrium distance is $\sim$0.25 *a*. The micrographs in Figure 14.5 obtained in a model experiment using Cu-coated W spheres demonstrate and confirm that the particles can be separated as the wetting angle increases.[2] In addition, under a high wetting angle, liquid can be agglomerated locally[2] and densification hardly occurs in real systems.

15

GRAIN SHAPE AND GRAIN GROWTH IN A LIQUID MATRIX

15.1 CAPILLARY PHENOMENA IN A BINARY TWO-PHASE SYSTEM[21]

The growth of grains in a liquid matrix is usually explained by the LSW theory.[17,18] Because of the capillary pressure exerted on a particle, the activity of the atoms in the particle and, hence, its solubility in the matrix increases as the particle size decreases. Therefore, the atoms dissolved in the matrix from small particles are transported to large particles, resulting in growth of large grains. To analyse this phenomenon quantitatively, it is first necessary to understand the capillarity in multi-component systems as well as that in single component systems (Section 2.3).

Consider an α solid solution sphere in a β matrix for a system shown in Figure 15.1(a). If α and β are incompressible and are at a constant temperature,

$$\sum_{i=1} (\mu_i^\alpha - \mu_i^\beta) dn_i = 0 \tag{15.1}$$

at equilibrium. Here, μ_i^α and μ_i^β are, respectively, the chemical potential of species i in the α and β phases under pressures P^α and P^β. For the α and β solutions, dn_i is independent, and hence

$$\mu_i^\alpha = \mu_i^\beta \tag{15.2}$$

and

$$P^\alpha - P^\beta = \gamma K \tag{15.3}$$

where γ is the interfacial energy between the particle and the matrix (if the matrix is a liquid, γ is γ_{sl}) and K the mean curvature.

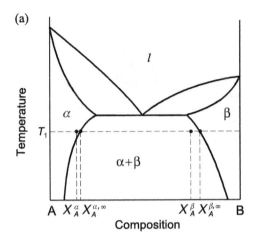

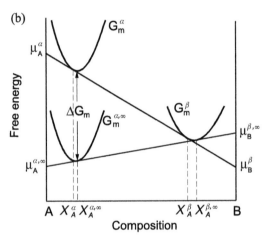

Figure 15.I. (a) Typical phase diagram showing limited solubility of $X_A^{\alpha,\infty}$ and $X_B^{\beta,\infty}$ at temperature T_I and (b) schematic of the molar free energy versus composition at the temperature for α precipitates with a flat interface ($K=0$) and with a finite radius of curvature ($K \neq 0$).

Similar to the case of a single component system, Eq. (15.2) can be expressed in terms of bulk chemical potentials and interfacial energy as

$$\mu_i^{\alpha}(T, P^{\beta}) - \mu_i^{\beta}(T, P^{\beta}) + \overline{V}_i^{\alpha}\gamma K = 0 \qquad (15.4)$$

where $\overline{V}_i^{\alpha}$ is the partial molar volume of i in the α phase, $(\partial \mu_i^{\alpha}/\partial P)_T$. For a two-component system, Eq. (15.4) becomes

$$\mu_B^{\alpha} - \mu_B^{\beta} + \overline{V}_B^{\alpha}\gamma K = 0 \qquad (15.5a)$$

and

$$\mu_A^\alpha - \mu_A^\beta + \overline{V_A}^\alpha \gamma K = 0 \tag{15.5b}$$

where A is the solute in the matrix and B the solvent. For $K=0$,

$$\mu_B^{\alpha,\infty} = \mu_B^{\beta,\infty} \quad \text{and} \quad \mu_A^{\alpha,\infty} = \mu_A^{\beta,\infty} \tag{15.6}$$

Since

$$\mu_B^\alpha - \mu_B^{\alpha,\infty} = RT\ln(a_B^\alpha / a_B^{\alpha,\infty}) \tag{15.7}$$

$$\ln\frac{v_B^\alpha v_B^{\beta,\infty}}{v_B^{\alpha,\infty} v_B^\beta} + \ln\frac{1 - X_A^\alpha}{1 - X_A^{\alpha,\infty}}\frac{1 - X_A^{\beta,\infty}}{1 - X_A^\beta} = -\frac{\overline{V_B}^\alpha \gamma K}{RT} \tag{15.8a}$$

where X_A^α and X_A^β are, respectively, the mole fractions of A in the α and β phases, and v_B the activity coefficient of B. Similarly,

$$\ln\frac{v_A^\alpha v_A^{\beta,\infty}}{v_A^{\alpha,\infty} v_A^\beta} + \ln\frac{X_A^\alpha X_A^{\beta,\infty}}{X_A^{\alpha,\infty} X_A^\beta} = -\frac{\overline{V_A}^\alpha \gamma K}{RT} \tag{15.8b}$$

In general, v is a function of X but it can be considered to be a constant for a small variation in X. Since the value of Eq. (15.8) is much smaller than 1 for a particle with a radius greater than 0.1 μm at a usual liquid phase sintering temperature, Eq. (15.8) can be written as

$$\frac{1 - X_A^\beta}{1 - X_A^\alpha}\frac{1 - X_A^{\alpha,\infty}}{1 - X_A^{\beta,\infty}} = \exp\left(\frac{\overline{V_B}^\alpha \gamma K}{RT}\right) \approx 1 + \frac{\overline{V_B}^\alpha \gamma K}{RT} \tag{15.9a}$$

and

$$\frac{X_A^{\alpha,\infty} X_A^\beta}{X_A^\alpha X_A^{\beta,\infty}} \approx 1 + \frac{\overline{V_A}^\alpha \gamma K}{RT} \tag{15.9b}$$

Multiplying $(1-X_A^\alpha)$ and X_A^α with Eq. (15.9a) and Eq. (15.9b), respectively, and adding them gives

$$X_A^\beta = X_A^{\beta,\infty}\left(1 + \frac{1 - X_A^{\beta,\infty}}{X_A^{\alpha,\infty} - X_A^{\beta,\infty}}\frac{\gamma V^\alpha K}{RT}\right) \tag{15.10a}$$

where V^α is the molar volume of $\alpha [= (1 - X_A^\alpha)\overline{V_B}^\alpha + X_A^\alpha \overline{V_A}^\alpha]$. A similar procedure gives

$$X_A^\alpha = X_A^{\alpha,\infty}\left(1 + \frac{1 - X_A^{\alpha,\infty}}{X_A^{\alpha,\infty} - X_A^{\beta,\infty}}\frac{\gamma(V^\alpha)'K}{RT}\right) \tag{15.10b}$$

where $(V^\alpha)' = (1 - X_A^{\beta,\infty})\overline{V_B}^\alpha + X_A^{\beta,\infty}\overline{V_A}^\alpha$. According to these equations, the increase in the solubility of A from an α sphere to the β matrix, ΔX_A^β, is proportional to the curvature of the particle. This result can be visualized in a free energy versus composition diagram as shown in Figure 15.1(b). The common tangent scheme in Figure 15.1(b) shows an increase in the solubility of A in β by an increase in the free energy of the particle, ΔG_m, due to its finite curvature.

Equation (15.10) was derived for systems where α and β are solutions of A and B. However, this equation can also be applied to systems where α is an intermediate compound, A_yB_x, and β is a liquid or a solid solution.[21] In this case, X_A^β is expressed as

$$X_A^\beta = X_A^{\beta,\infty}\left(1 + \frac{1 - X_A^{\beta,\infty}}{X_A^{\alpha,\infty} - X_A^{\beta,\infty}}\frac{\gamma V^\alpha K}{RT}\right) \tag{15.11}$$

In this equation, $X_A^\alpha = y/(x+y)$ and $V^\alpha = V_c/(x+y)$, where V_c is the molar volume of the compound.

15.2 LIFSHITZ–SLYOZOV–WAGNER (LSW) THEORY[17,18,22]

When particles of different sizes are dispersed in a liquid, material transport occurs from small to large grains because of the difference in solubility between the grains. Therefore, small grains dissolve and large grains grow further, and the average grain size increases, a phenomenon called 'Ostwald ripening'. Lifshitz and Slyozov[17] rigorously analysed the growth phenomenon controlled by diffusion of atoms in the matrix and Wagner[18] that controlled by atom diffusion and also by the reaction at the solid/liquid interface.

In the development of the LSW theory, a few essential conditions were assumed. The theory first considers an infinitely dispersed system where the solid volume fraction is theoretically zero. A more basic but not explicitly noted assumption is that the dissolution and precipitation of atoms occur with an equal probability for each atom regardless of the size and the crystallographic plane of the grains. In other words, the mobility of an interface is constant and independent of the driving force and crystallographic orientation of the solid surface. This assumption, however, is acceptable only for rounded grains with a rough interface.[23–25] (In this case grain growth is controlled by atom diffusion.) Under these assumptions, the theory predicts a constant (stationary) size distribution of grains, irrespective of the initial grain size distribution, and a simple kinetic equation in a stationary state: the cube (diffusion control) or the square (interface reaction control) of the average grain size being proportional to the holding time. Instead of the derivation of the LSW theory, which is not easy, this section will introduce an approximate but simple derivation of Greenwood.[22] (The growth controlled by interface reaction will also be described following the original Wagner theory, although the basic assumption of a constant interface mobility is not satisfied for the interface-reaction-controlled growth (see Section 15.2.3 and Section 15.4).)

Figure 15.2 schematically shows the concentration profiles of a solute in a liquid between a small and a large grain for diffusion-controlled (Figure 15.2(a)) and interface-reaction-controlled growth (Figure 15.2(b)). In the case of diffusion control where equilibrium at the interface is maintained, the concentrations of the solute at the interface of an α grain and liquid are defined as Eqs (15.10b) and (15.10a), respectively. The average solute concentration in the

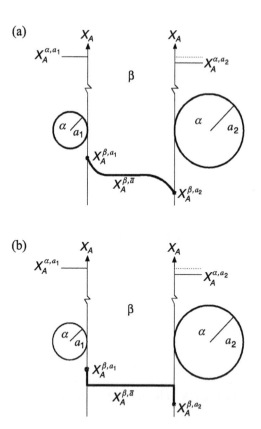

Figure 15.2. Schematic of solute concentration gradient in a matrix between two particles with different size a_1 and a_2: (a) diffusion control and (b) interface-reaction control.

liquid is the solute solubility of the average size grain, as will be shown later. For grains with an infinite separation, the concentration gradient of the solute at the solid/liquid interface of a grain with a radius of a is expressed as*

$$\left.\frac{dC}{dR}\right|_{R=a} = \frac{C_{\bar{a}} - C_a}{a} \tag{15.12}$$

where R is the distance from the centre of the grain concerned, $\bar{a}$ the radius of the average sized grain and $C_{\bar{a}}$ the solute solubility of the average sized grain. On the other hand, for reaction-controlled growth, there is essentially no gradient in solute concentration in the liquid and the solute concentration in the liquid is approximately the solute solubility of the average sized grain (see Section 15.2.2). This suggests that the kinetics of dissolution is the same as that of precipitation, which would hardly be justified. (See Section 15.2.3.)

15.2.1 Grain Growth Controlled by Diffusion

For diffusion-controlled growth, consider a material flux (moles/s) passing through an imaginary spherical surface in the liquid matrix a distance of R from a grain with a radius of a. This material flux must be equal to the rate of the mass change of the grain. If the α phase is pure A and the molar volumes of α and β are the same,

$$-\frac{4\pi R^2 D}{V_m}\frac{dC}{dR} = \frac{4\pi a^2}{V_m}\frac{da}{dt} \tag{15.13}$$

where D is the diffusion coefficient of A in the liquid. Assuming that the instantaneous rate of grain size change da/dt is constant, the integration of Eq. (15.13) gives

$$\frac{da}{dt} = -\frac{D(C_a - C_{\bar{a}})}{a} \tag{15.14}$$

Since $C_a = C_\infty(1 + 2\gamma V_m/RTa)$,

$$\frac{da}{dt} = \frac{2D\gamma C_\infty V_m}{RTa}\left(\frac{1}{\bar{a}} - \frac{1}{a}\right) \tag{15.15}$$

Figure 15.3 plots the variation of growth rate with grain size (Eq. (15.15)). The grains smaller than the average sized grain dissolve and those larger than the average sized grain grow. The maximum growth rate appears at $a = 2\bar{a}$.

---

*According to the notation in Section 15.1, C is X_A^β. Hereafter, we denote X_A^β as C for convenience.

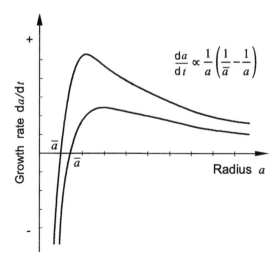

Figure 15.3. Variation of particle growth rate da/dt for diffusion-controlled growth (Eq. (15.15)) for two different values of the mean radius $\bar{a}$.

However, as the annealing time increases, the particles that have been growing can begin to dissolve and theoretically only the largest grain present at the beginning remains to the end.

The change in average grain size with annealing time was predicted by the LSW theory. However, since Eq. (15.15) holds for each grain, we may just assume that d$\bar{a}$/d$t \approx$ (da/dt)$_{\text{max}}$.[22] Then, the integration of Eq. (15.15) gives $\bar{a}_t^3 - \bar{a}_0^3 = (3/2)(D\gamma C_\infty V_m/RT)t$. This equation is the same as the LSW equation,

$$\bar{a}_t^3 - \bar{a}_0^3 = \frac{8}{9}\frac{D\gamma C_\infty V_m}{RT}t \qquad (15.16)$$

except for the proportionality constant. According to the LSW theory, irrespective of the initial grain size distribution, the grain size distribution becomes stationary on extended annealing with $a_{\text{max}} = 1.5\bar{a}$, as shown in Figure 15.4. In the stationary distribution the maximum frequency is at $a = 1.135\bar{a}$ and the average grain size $\bar{a}$ is the same as the critical grain size a^* of a grain that neither dissolves nor grows, i.e. da/d$t = 0$.

15.2.2 Grain Growth Controlled by Interface Reaction

Assuming that the reaction (atom attachment and detachment) at the grain interface is linearly proportional to its driving force, Wagner derived a kinetic equation of interface-reaction-controlled grain growth and a stationary size distribution of grains. This assumption is questionable (see Section 15.2.3) and

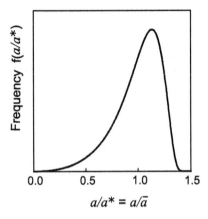

Figure 15.4. Stationary distribution of normalized particle sizes for diffusion-controlled growth.

the result is not applicable for explaining the real phenomena that occur by reaction control. In this section, however, Wagner's result is introduced merely as a reference. Under the reaction assumption of Wagner, the grain growth rate, da/dt, is expressed as

$$\frac{da}{dt} = K(C_{\bar{a}} - C_a) \tag{15.17}$$

where K is a constant including the interface mobility. Comparing this equation with Eq. (15.14), it is seen that, with a reduction in grain size, the controlling mechanism of grain growth for a system can change from diffusion control to reaction control. However, the feasibility of the change in controlling mechanisms in real systems is unknown because the above comparison was made under the questionable assumption that the interface mobility is constant for both diffusion and reaction control.

From Eq. (15.17), since

$$\frac{da}{dt} = \frac{2K\gamma C_\infty V_m}{RT}\left(\frac{1}{\bar{a}} - \frac{1}{a}\right) \tag{15.18}$$

the growth rate changes with grain size in a manner shown in Figure 15.5. Using an assumption similar to that for diffusion control for an average growth rate, we can easily derive a kinetic equation, identical to that of Wagner except for the proportionality constant. The original equation of Wagner is

$$\bar{a}_t^2 - \bar{a}_o^2 = \frac{64}{81}\frac{K\gamma C_\infty V_m}{RT}t \tag{15.19}$$

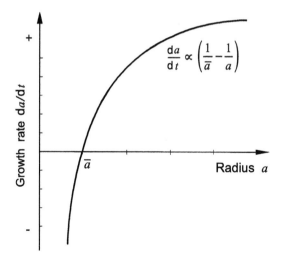

Figure 15.5. Variation of particle growth rate da/dt for interface reaction-controlled growth (Eq. (15.18)).

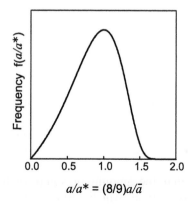

Figure 15.6. Stationary distribution of normalized particle sizes for interface reaction-controlled growth.

for a steady state. Figure 15.6 shows the steady-state distribution of grain size. Unlike the case of diffusion control, the average grain size $\bar{a}$ in reaction-controlled growth is not the same as the critical grain size a^*, $\bar{a} = (8/9)a^*$. The maximum grain size is predicted to be $2a^*[= (9/4)\bar{a}]$.

15.2.3 Verification and Limitations of LSW Theory

To analyse the grain growth in a real system using the LSW theory, the validity of the basic assumptions that are involved in the derivation of the theory has first to be confirmed. The most critical assumption is that of constant interface (grain/matrix interface) mobility under all conditions, including changes in driving force and crystallographic plane of the solid. Inferring that the rates of atom attachment and detachment are the same under any condition, this assumption is justified only for round grains with atomically rough interfaces. For faceted grains with flat interfaces the assumption does not hold.[23-28] When the driving force for grain growth is low, as in Ostwald ripening, faceted grains can grow only by either two-dimensional nucleation on a flat surface or with the assistance of defects, such as screw dislocations or twins.[23-31] These growth mechanisms of faceted grains mean that growth is governed by interface reaction, which is basically different from Wagner's assumption for reaction-controlled Ostwald ripening. The shape of a faceted grain itself demonstrates that even under a constant driving force, the interface mobility varies with crystallographic planes of the grain. Therefore, the Ostwald ripening of faceted grains cannot be explained by Wagner's theory for interface-reaction-controlled growth. Recently, some attempts at numerical calculation[32,33] have been made to explain the Ostwald ripening of faceted grains whose interface mobility is variable with driving force and crystallographic orientation.

Another assumption is that of an infinite dispersion of grains. Under this assumption, whether a grain is growing or dissolving is determined by the size of the grain relative to the critical grain size (in the case of diffusion control, the average size) and its rate is given by Eqs (15.15) and (15.18). In other words, each grain interacts with a critical sized grain in the LSW theory. However, in real systems with a finite volume of solid grains, grain growth rate and grain size distribution change with solid volume fraction (see below) in diffusion-controlled growth. In addition, with increasing solid volume fraction, the shape of a grain can be affected by the nearest neighbour grains.[15,34]

It is also assumed that the amount of solute in the matrix is constant (invariable) with annealing time, which results from the infinite dispersion assumption. In real systems, however, the solute solubility in the matrix should decrease with grain growth, which results in an increase in solid volume fraction. (This result is easily understandable by reference to a phase diagram, such as Figure 15.1(a).) The assumption, however, is not critical in real systems where the grain size is, in general, larger than a micron and the solubility change with grain size is insignificant.

Another assumption concerns the rate controlling species. In the LSW theory, the species that controls the grain growth is the solute in the matrix. For compounds, however, the controlling species may vary.[35,36] Maintenance of the correct chemical composition of the grains also requires a flux coupling of the species in the matrix. This composition constraint can be met by an effective diffusion coefficient as in the case of the diffusion in an ionic compound (see Section 13.1).

In addition to his theory of diffusion-controlled growth, Wagner further developed a theory of reaction-controlled growth. However, this theory is not acceptable for real systems, as explained above. Physically and by definition, a fundamental difference between growth controlled by diffusion and growth controlled by interface reaction is their dependence on the volume fraction of the matrix. In the case of diffusion control, the growth rate must decrease as the matrix volume fraction increases. However, for reaction-controlled growth, the rate is theoretically independent of matrix volume fraction. Based on this concept, some experimental tests have been made. In the case of SiAlON grains in an oxynitride glass, the growth was independent of the liquid volume fraction, implying that growth is controlled by interface reaction.[37,38] A previous study on TaC-Co alloys also showed that the growth rate of faceted TaC grains was independent of the Co matrix fraction.[39] The independence of grain growth rate on liquid volume fraction, however, should be valid only for specific ranges of liquid volume fraction and driving force for grain growth. (See Section 15.4.)

The kinetic equations and grain size distributions of the LSW theory are concerned with growth in a steady state (time-independent distribution of grains). If a steady state growth condition is not satisfied due to, for example, volume fraction change with annealing time (such as in the initial stage of precipitation), non-constant solute concentration in the grains (such as in SiAlON ceramics) and non-steady state distribution of grains (such as in the initial stage of grain growth), care must be taken when using the theoretical equations in analysing grain growth data. Therefore, for data analysis based on the LSW theory, steady state growth in the system concerned must first be confirmed. In a number of previous investigations on the growth of precipitates or grains in a matrix, kinetic analyses have been made without such confirmation.

Since the volume fraction of grains is not zero in reality, unlike that in the LSW theory, the growth kinetics is affected by the volume fraction when growth is governed by diffusion of atoms in the liquid (diffusion control). A number of theoretical and experimental investigations have been made on the volume fraction effect on grain growth kinetics and grain size distribution.[40–51] According to the results, the growth equation is basically the same as Eq. (15.16) but the proportionality constant increases as the volume fraction of grains increases. As for grain size distribution, an increase in grain volume fraction broadens the distribution. For systems where a mean field concept can be applied, as in the LSW theory, the grain size distribution is predicted to be that of Wagner's reaction controlled growth.[41] However, for systems where the local environment affects the growth of the grain concerned, i.e. a communicating neighbour concept is applicable, the distribution is predicted to be different to the LSW distributions. In reality, since the shape of an individual grain is affected by its neighbour grains,[15,34] the communicating neighbour concept should be more realistic. The size distribution of grains in a liquid matrix has been measured in a number of systems.[42,45–47,49] The Rayleigh distribution fraction was recently suggested to best fit the measured grain size distributions.[51]

To measure the microstructural characteristics of liquid phase sintered materials, such as the grain volume fraction, average grain size and grain size distribution, quantitative metallographic techniques including point, lineal and areal analyses are utilized.[52–55] Details of these analyses are well documented in many references and have been developed as computer programs. To estimate the growth kinetics, however, a simple measurement of the maximum grain size as a function of annealing time is sufficient if the grain size distribution does not vary with the annealing time.

15.3 GRAIN SHAPE IN A LIQUID

When a grain is immersed for a long period of time in a liquid which is chemically in equilibrium with the grain, the grain exhibits a shape with the minimum interfacial energy, an equilibrium shape. On the other hand, for a system with many grains with different sizes, dissolving small grains exhibit shrinkage shapes and growing large grains exhibit growth shapes.

If external forces, including gravity and surface stresses, can be neglected, the equilibrium shape of a crystal in a matrix is determined by the minimization of the total interfacial energy, $\sum_i \gamma_i A_i$,[56] where γ_i is the specific interfacial energy of the crystallographic plane i and A_i its area. In other words, when a shape change under constant temperature, constant volume and constant composition results in an increase in the total interfacial energy, i.e.

$$\delta \left(\sum_{i=0}^{m} \gamma_i A_i \right) > 0 \tag{15.20}$$

the shape of the original crystal was an equilibrium one.[56] Therefore, if the interfacial energy is isotropic, as in the case of a liquid drop, the equilibrium shape is a sphere. Otherwise, the crystal can have a shape with edges, corners and faceted planes. Knowing a simple relationship between the surface energy of a crystal plane and the vertical distance from the crystal centre to the crystal surface plane, Wulff[57] developed a theorem from Eq. (15.20), which predicts the equilibrium shape of a crystal from the surface energies of the crystal and their crystallographic orientations. Later, Herring[57] showed that the Wulff theorem holds for crystals with partially rounded surfaces.

The Wulff theorem can be derived from the equilibrium condition of an isolated system containing a single crystal, i.e. the minimum free energy condition of the system.[58,59] Under equilibrium at a constant temperature, an infinitesimal change in Helmholtz free energy of the system, dF, is expressed as,*

$$dF = \sum_i \gamma_i dA_i + \left(\frac{\partial F}{\partial n^c}\right)_{T,V} dn^c + \left(\frac{\partial F}{\partial V^c}\right)_{T,V} dV^c$$
$$+ \left(\frac{\partial F}{\partial n^s}\right)_{T,V} dn^s + \left(\frac{\partial F}{\partial V^s}\right)_{T,V} dV^s = 0 \qquad (15.21)$$

Here, n represents the number of moles, V the volume, and c and s the single crystal and the surrounding phase, respectively. Equation (15.21) is rewritten as

$$\sum_i \gamma_i dA_i + (\mu^c - \mu^s)dn^c - (P^c - P^s)dV^c = 0 \qquad (15.22)$$

Let h_i be the distance perpendicular from the crystal centre to the plane i, $V^c = 1/3(\sum_i A_i h_i)$ and

$$dV^c = \frac{1}{2}\sum_i h_i dA_i \qquad (15.23)$$

$$\therefore \sum_i \left[\gamma_i - \frac{h_i}{2}(P^c - P^s)\right]dA_i + (\mu^c - \mu^s)dn^c = 0 \qquad (15.24)$$

If dA_i and dn^c are independent of each other,

$$P^c - P^s = \frac{2\gamma_i}{h_i} \qquad (15.25a)$$

and

$$\mu^c = \mu^s \qquad (15.25b)$$

At equilibrium, $(P^c - P^s)$ is constant and

$$\frac{2\gamma_1}{h_1} = \frac{2\gamma_2}{h_2} = \cdots = \frac{2\gamma_i}{h_i} \equiv K_w \qquad (15.26)$$

where K_w is the Wulff constant (Wulff theorem). The chemical potential μ of the atoms in the crystal is then

$$\mu = \mu^o + \frac{2\gamma_i V_m}{h_i} \qquad (15.27)$$

of which the form is similar to that for a sphere.

---

*Use of Helmholtz free energy allows us to avoid consideration of the volume change with pressure.

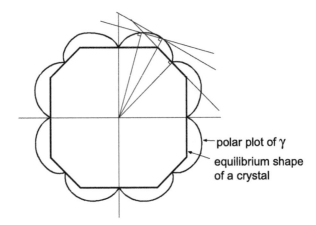

polar plot of γ

equilibrium shape
of a crystal

Figure 15.7. Polar plot of surface energy γ and Wulff construction for an equilibrium crystal shape.

Figure 15.7 shows an example of a Wulff construction which allows us to determine an equilibrium shape of a crystal from its γ-plot (a polar diagram showing a variation of surface energy with the crystallographic orientation). The cusps in the γ plot appear when the internal energy (binding energy) of atoms is large compared with the entropy contribution to free energy. The γ plot at 0K can easily be constructed by considering the binding energy of atoms without the entropy contribution;[60] the equilibrium shape of crystals at 0K is fully faceted.

To determine the equilibrium shape from a γ-plot, connect all of the γ points in the polar diagram to the centre of the diagram, and construct perpendicular planes (Wulff planes) on the γ points. Then, the minimum volume enveloped by the Wulff planes exhibits the equilibrium shape of the crystal. This is because the surface energy of a plane is proportional to the distance from the centre to the γ point, and, therefore, the total surface energy is proportional to the volume enveloped by Wulff planes. When a minimum cusp is present in a γ-plot, a facet plane appears. For a point and a line of energy maximum in a γ-plot, a corner and an edge appear in the equilibrium shape.

To observe the equilibrium shape of a crystal experimentally, it is necessary to confirm that the crystal one wants to observe exhibits such a shape. Since the equilibrium shape of a crystal in a matrix is the same as the equilibrium shape of the matrix entrapped within the crystal, the equilibrium crystal shape can be determined by observing the shape of the entrapped matrix.[61–65] If a number of randomly oriented grains exhibit an equilibrium shape, the shape can also be determined stereographically from the crystal plane orientations of the grains and the directions of the grain interfaces on a planar section that can be obtained by the electron backscattered diffraction technique.[66,67] For metals for which the anisotropy in interfacial energy is low at their processing

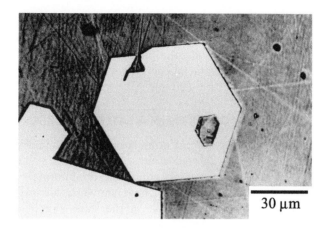

Figure 15.8. Microstructure showing $MgAl_2O_4$ spinel grains in a CaMgSiAlO glass matrix and an entrapped glass pocket within a grain.[62]

temperatures, the equilibrium shape is usually a sphere or rounded with sharp corners and edges. On the other hand, ceramics usually exhibit high anisotropy in interfacial energy and their equilibrium shape is often a polyhedron.[62–65] Figure 15.8[62] shows an $MgAl_2O_4$ spinel grain in a CaMgSiAlO glass and a glass drop entrapped within the grain. From the shapes of the entrapped second phase on several cross-sections as well as the three-dimensional shape observed after a complete elimination of the matrix, one can determine the equilibrium shape of the grain. In the case of the $MgAl_2O_4$ spinel in the CaMgSiAlO glass, the equilibrium shape was determined to be an octahedron.[62]

The equilibrium shape of a crystal varies with thermodynamic parameters, such as chemical composition and temperature, because the interfacial energy is a function of these parameters. In many cases, the anisotropy in interfacial energy decreases as impurities segregate at the interface. An increase in temperature results in the reduction of energy anisotropy because of an increased vacancy concentration and thus an increased contribution of entropy.[68] Similarly, an increase in vacancy concentration by other means, for example oxygen partial pressure change, also reduces the interfacial energy anisotropy.[69] A faceted interface can, therefore, become rough (interface roughening[23]) with dopant addition,[69–71] temperature increase[71,72] or oxygen partial pressure change.[69]

During liquid phase sintering, the grains grow, mostly, under chemical equilibrium. If the solid/liquid interfacial energy is isotropic, the shapes of growing and dissolving grains (G and D shapes) are spherical (rounded) because the rates of growth and dissolution are independent of the crystallographic orientation. However, if the rates vary with the crystallographic orientations, the shapes become round-edged or simple polyhedra. In the case of simple polyhedra, the rates are usually controlled by interfacial reaction and the shapes can basically be explained by the Frank theory.[73,74] For a growth shape, slowly growing planes appear and they usually have low interfacial energies. On the other hand, a dissolution shape consists of fast dissolving planes. However, when the interfacial energy anisotropy is not very high, the dissolution shape tends to be rounded and spherical.[71,72]

15.4 ABNORMAL GRAIN GROWTH IN A LIQUID MATRIX

A number of examples of abnormal grain growth in a liquid matrix exist; however, all of them involve faceted grains.[27,69,75–80] This suggests that the faceting of grain interfaces is a necessary condition for abnormal grain growth. In particular, recent investigations[69,79,80] have shown a correlation between grain shape and grain growth behaviour in the same system. For faceted grains in the systems $SrTiO_3$,[69] SiC,[79] and $BaTiO_3$,[80] abnormal grain growth occurred. However, when interface roughening was induced, i.e. the grain shape became rounded by any means, for example, increasing sintering temperature, changing sintering atmosphere (oxygen partial pressure) or adding dopants, the growth behaviour became normal. These observations in the same system strongly support the theory that faceting of grains can induce abnormal grain growth.

Theoretical explanations and analyses of abnormal grain growth in a liquid matrix have recently been attempted.[27,28,32,33] These are based on previous experiments and theories of crystal growth from a melt. According to previous crystal growth studies,[23,24,81] a faceted crystal grows either with the assistance of surface defects, such as screw dislocations and twin boundaries, or by two-dimensional nucleation and growth without surface defects, as shown schematically in Figure 15.9. For crystals with a rough interface, growth is controlled by diffusion and growth rate is linearly proportional to driving force, i.e. the interface mobility is constant. On the other hand, crystals with a faceted interface grow by interface-reaction control if the diffusion of atoms in the matrix is sufficiently fast; the growth rate is not a linear function of driving force, and the interface mobility which is a function of driving force also varies with the crystallographic plane of the interface. According to crystal growth theories,[23,24,81] the growth rate (velocity) v of a facet is expressed as

$$v = A_1 \frac{(\Delta G)^2}{\varepsilon} \tanh\left(\frac{\varepsilon}{\Delta G}\right) \tag{15.28}$$

for screw dislocation-assisted growth. Here, A_1 is a constant including material and physical constants, ΔG the driving force for growth, and ε the step free energy of the crystal (also called edge energy). On the other hand, for two-dimensional nucleation and growth, the growth rate is expressed as

$$v \propto A_2 \left(\frac{\Delta G}{T}\right)^n \exp\left(-\frac{A_3 \varepsilon^2}{T \Delta G}\right) \tag{15.29}$$

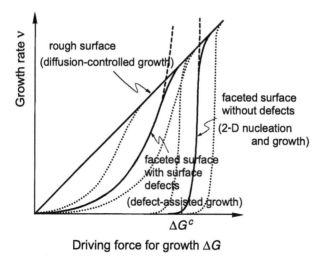

Figure 15.9. Schematic showing the variation of growth rate with driving force for various growth modes.[27] The dotted lines depict the growth rates for different crystallographic planes of a faceted crystal.

where A_2 and A_3 are constants including material and physical constants, T is the interface temperature, and n is an exponent with value 1/2 for mononucleation and growth (MNG) and 5/6 for polynucleation and growth (PNG). A_2 includes the area of the interface for MNG but does not for PNG.

The growth equations (Eqs (15.28) and (15.29)) indicate that the growth rate is basically a parabolic function of the driving force for defect-assisted growth and an exponential function for two-dimensional nucleation and growth, as shown schematically in Figure 15.9. These results suggest that there is a critical driving force for significant growth of the crystal, in particular, for growth without surface defects. The critical driving force ΔG^c varies with the step free energy of a two-dimensional nucleus which is, in turn, affected by temperature and dopant. With a reduction of step free energy, ΔG^c decreases.

Noticing that the growth of grains in a liquid matrix is identical to the growth of a single crystal from a melt, Park et al.[28] suggested that significant growth of faceted grains is possible only when the driving force for the growth is larger than a critical value, ΔG^c, as shown in Figure 15.9. The driving force for the growth of an individual grain comes from the difference in size between the grain concerned and the grain with a zero rate of growth. For a faceted grain with an effective radius of a to have significant growth,

$$\Delta G^c \leq 2\gamma V_m \left(\frac{1}{a^*} - \frac{1}{a} \right) \tag{15.30}$$

must be satisfied, where a^* is the critical radius of grains with a zero rate of growth $(da/dt = 0)$ at the moment of concern. To satisfy Eq. (15.30) for significant growth, a^* must be smaller than a critical value and a must be larger than a certain value. Significant growth will occur for grains with driving forces larger than ΔG^c while growth will be practically absent for those with driving forces smaller than ΔG^c. The result of such a selective growth of grains must be a bimodal distribution of grains in size, showing abnormal grain growth. For a given system under a constant temperature, there are numerous relative values of a^* and a that satisfy Eq. (15.30). With decreasing average grain size, the number of grains that satisfy Eq. (15.30) increases, while no grain can satisfy the equation if the average grain size exceeds a certain value. This conclusion suggests that no abnormal grain growth results for coarse grains even if the grains are faceted. This prediction is consistent with experimental observations in WC-Co[28,75] and BaTiO$_3$.[80]* The presence of a critical driving force for significant growth of faceted grains can also be found in a seeding experiment.[28] Large seed grains with high driving forces for growth induced abnormal grain growth in WC-Co compacts where no abnormal grain growth occurred without seeds.[28] However, in systems with rounded grains, seeding did not induce any abnormal grain growth.[82] These theoretical and experimental considerations[23,24,27,28,75,80–82] confirm that abnormal grain growth can occur only when the grains in a matrix are faceted and small and grow by interface reaction control.

As shown in Figure 15.9, the growth rate of a faceted surface follows Eq. (15.28) or Eq. (15.29) if the growth rate is lower than that of a rough surface while it follows a diffusion-controlled growth equation, basically Eq. (15.15), if the growth rate exceeds that of a rough surface. In other words, the growth rate is determined by the slower process in the two serial processes, diffusion and interface reaction. When diffusion of atoms is slower than interface reaction, the growth of a facet is also governed by diffusion. The growth rate of a facet v over a wide range of values for driving force is then expressed as

$$v = \frac{v_R v_D}{v_R + v_D} \tag{15.31}$$

where v_R is the growth rate controlled by interface reaction (Eqs (15.28) or (15.29)) and v_D that controlled by diffusion (Eq. 15.15). Under a high driving force and with a high growth rate, kinetic roughening of facets can occur[83] showing an apparently rounded shape in the microscale with well developed facets in the nanoscale.[69] Since the growth rate of a faceted grain under a high

---

*A similar grain size effect observed in BaTiO$_3$ without a liquid phase[80] supports the explanation of abnormal grain growth in single-phase systems (Section 9.2.1), which is based on the non-linear mobility of a faceted grain boundary with the driving force, as in the case of a solid/liquid interface (Figure 15.9).

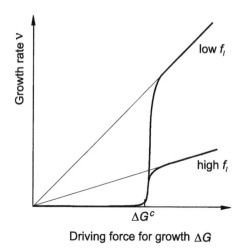

Figure 15.10. Schematic showing the effect of liquid volume fraction f_l on the growth rate of a facet.

driving force is governed by diffusion, the growth behaviour of faceted grains in a liquid matrix is expected to be affected by the volume fraction of the liquid. Figure 15.10 shows schematically the variation of the growth rate of a facet with driving force in a liquid with a low and a high volume fraction. For a low volume fraction with a high diffusion rate, the growth rates of faceted grains are very different, depending on the driving forces for their growth: very high rates for grains with driving forces larger than ΔG^c and very low rates for those with driving forces smaller than ΔG^c. As the volume fraction of liquid increases, the growth rate controlled by diffusion decreases.[40–51] Then, the difference in growth rate between grains with larger and smaller driving forces than ΔG^c is reduced. This consideration suggests that for a given system with faceted grains where abnormal grain growth occurs in a liquid with a low liquid volume fraction, less abnormal or apparently normal grain growth behaviour can be exhibited for a high liquid volume fraction.[84] This expectation of a liquid volume fraction effect is, indeed, observed in some systems with faceted grains such as NbC-Ni[85] and TiC-Ni.[86]

As Eqs (15.28) and (15.29) show, the critical driving force ΔG^c decreases with a reduction of step free energy. The step free energy varies considerably with temperature and dopant. Since step free energy (in turn, activation energy for two-dimensional nucleation) decreases as the temperature increases,[87–89] ΔG^c decreases with a temperature increase. With a reduction of step free energy and thus a reduction of ΔG^c, the probability of having large abnormal grains is expected to increase in the same system if the reduction is not too high to induce pseudo-normal grain growth. (If the step free energy is zero, diffusion-controlled growth occurs and normal grain growth results.) In fact, abnormal grain growth is facilitated with temperature increase in the Si_3N_4 system.[90,91]

Simulations on the effect of step free energy on grain growth behaviour have also been made.[32,92,93] Figure 15.11 shows the result of a Monte Carlo simulation made by Cho.[93] For the simulation, Cho assumed that the grain network was a set of grains with a Gaussian size distribution (standard deviation of 0.1) located on vertices of a two-dimensional square lattice. Deterministic rate equations, Eq. (15.15) for v_D and Eq. (15.29) for v_R, were

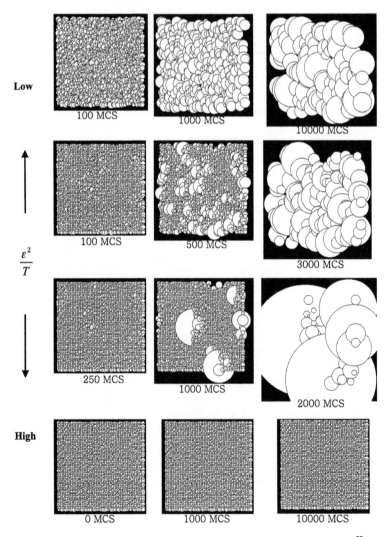

Figure 15.11. Simulated developments of microstructure with step free energy ε.[93] The initial size distribution was assumed to be Gaussian.

used to calculate the mass transport between grains using Eq. (15.31). As Figure 15.11 shows, the growth behaviour depends critically on the step free energy, more specifically ε^2/T. For low values of ε^2/T, the growth behaviour is apparently normal. As the ε^2/T value increases, abnormal grain growth behaviour appears. The abnormal grain growth behaviour becomes more distinct for larger ε^2/T because two-dimensional nucleation is more difficult and only a few large grains with driving forces larger than ΔG^c can grow abnormally. With the reduction of the number of abnormal grains, the size of the abnormal grains increases for the same period of annealing time. However, for very large ε^2/T, two-dimensional nucleation on any grain surface can be suppressed and apparently no growth can result, as shown in Figure 15.11. A similar dependence of grain growth behaviour on step free energy was also predicted by numerical analysis based on the mean field approach for faceted crystals dispersed in a liquid[32] or on a substrate.[92]

In recent years, utilization of abnormal grain growth behaviour in poly-crystals for materials processing has been intensively studied[94–99] (e.g. the fabrication of textured polycrystals[94–96] and single crystals[97–99]). Textured polycrystals can be made by embedding seeds (or templates) in fine matrix grains and annealing them at high temperatures. During the growth of the seeds, chemical reactions may occur if the seeds are not in chemical equilibrium with matrix grains and crystallographically oriented large faceted grains can form, a phenomenon called (reactive) templated grain growth.

Another example is the fabrication of single crystals from powder compacts with or without seeds.[97–99] Without seeds, it is necessary to control the

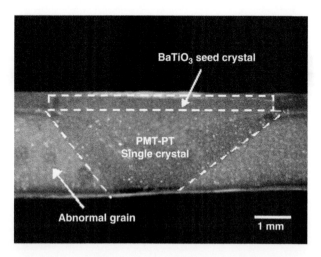

Figure 15.12. Growth of a $Pb(Mg_{1/3}Nb_{2/3})O_3$–$35PbTiO_3$ (mol%) (PMN-PT) single crystal into a fine (PMN-PT)-$2Li_2O$-$6PbO$ (mol%) matrix from a $BaTiO_3$ seed crystal during anneal-ing at $1200°C$ for 50 h.[99] Some abnormal grains are visible in the matrix.

nucleation of abnormally large grains in the compacts. Theoretically, if only a large grain is available in a powder compact, a single crystal with a volume of the powder material can be made. Single crystals can also be fabricated from powder compacts using single crystal substrates or seeds. (An example is shown in Figure 15.12.[99]) In this case, too, suppression of the formation of large abnormal grains in powder compacts is essential. Advantages of this technique over the conventional techniques of single crystal growth from a melt or vapour include simplicity in the technique itself, easy control of chemistry and little fluctuation in chemical composition of the crystals made. On the other hand, suppression of pore entrapment within growing crystals is critical to prepare fully dense single crystals. It is now commercially possible to produce almost pore-free single crystals of $BaTiO_3$ and PMN-PT ($Pb(Mg_{1/3}Nb_{2/3})O_3$-$PbTiO_3$) with a size of a few centimetres.[100] (This technique is referred to as the solid-state single crystal growth (SSCG) technique.) Since this technique utilizes the abnormal grain growth in seeded polycrystals, it would theoretically be applicable to any system with faceted interfaces that frequently exhibits abnormal grain growth behaviour.

16

DENSIFICATION MODELS AND THEORIES

So far, two models and theories have been developed for explaining densification during liquid phase sintering. One is the classical three-stage model[6] and theory[7] and the other the pore filling model[11,13] and theory.[8]

16.1 CLASSICAL MODEL AND THEORY

The classical model[6] and theory[7] of liquid phase sintering asserts that liquid phase sintering can be divided into three stages:

(i)   particle rearrangement by liquid flow,
(ii) contact flattening by a solution/reprecipitation (also called dissolution/precipitation) process, and
(iii) solid state sintering.

16.1.1 Initial Stage: Particle Rearrangement

Kingery[7] considered that the initial stage densification occurred via particle rearrangement during liquid flow immediately after the formation of a liquid. Referring to the kinetic equation of viscous flow in solid state sintering, he suggested that shrinkage occurred in accordance with the equation

$$\frac{\Delta l}{l} = \frac{1}{3}\frac{\Delta V}{V} \propto t^{1+y} \tag{16.1}$$

where t is the sintering time and y a constant smaller than unity that reflects the modification of densification kinetics due to an increase in resistance against viscous flow during particle rearrangement and an increase in driving force by pore size reduction. Kingery further proposed that the absolute increase in sintered density was linearly proportional to the volume fraction of liquid and that full densification was possible when the liquid volume fraction exceeds a certain value.

However, justification for the use of Eq. (16.1) in this way is uncertain, and in reality the proposal is merely a suggestion without any experimental basis. Whether viscous flow occurs or not at the beginning of liquid phase sintering depends strongly on the dihedral angle between the solid grains and the volume fraction of liquid. Of the two parameters, dihedral angle must be the prime. When this dihedral angle is greater than 0° and a solid skeleton forms during heating to the liquid phase sintering temperature as in W-Ni-Fe alloys, no viscous flow of particles and no particle rearrangement are expected.[5]

Viscous flow and particle rearrangement may occur only when the dihedral angle is 0°.[2,3,5] If the liquid volume fraction is high, viscous flow can occur, as shown in Figure 16.1 in the Mo-Ni system.[3] However, for a low liquid volume fraction, local rearrangement of particles must be predominant as

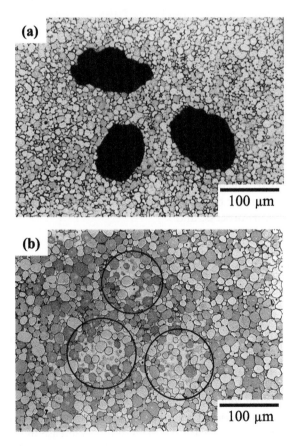

Figure 16.1. (a) Large pores of about 100 μm in size formed at large Ni particle sites after sintering a 92Mo(7 μm)-5Ni(1.5 μm)-3Ni(100 μm) specimen at 1460°C for 1 min and (b) liquid–grain pockets (circles) formed at large pore sites by grain–liquid mixture flow in the same specimen sintered at 1460°C for 3 min.[3.]

demonstrated recently by computer simulation.[4,101] The simulation shows that, while a slight shrinkage can be achieved, larger pores are formed by particle rearrangement, as in the case of solid state sintering.[102] This result suggests that particle rearrangement may not be beneficial for overall densification because the final densification is governed by the elimination of the largest pores in the compact.

16.1.2 Intermediate Stage: Contact Flattening

In the classical theory,[7] the intermediate stage is characterized by so-called contact flattening (a change in grain shape as the result of a solution/ reprecipitation process). The microstructure immediately after the particle rearrangement stage was assumed to consist of uniformly distributed grains and mono-sized pores in a liquid matrix. With this assumption, the microstructure can be simplified and represented as two particles, with liquid and a pore at the neck surface, as shown schematically in Figure 16.2. Kingery[7] further assumed no grain growth during densification, a soluble solid in a liquid, and presence of a liquid film between the particles, i.e. a dihedral angle of 0°.

Under these conditions, a compressive pressure* is exerted between the two particles due to the capillary pressure and the surface tension of the liquid. (See Section 14.2.) Because of this compressive pressure at the contact area, the chemical potential of atoms at the contact area is higher than elsewhere (i.e. at the surface of the neck region). Therefore, the solid at the contact area dissolves and the dissolved material is transported to the surface of the neck. As a result of this material transport, the contact area increases and the grain shape is accommodated (contact flattening). At the same time, the pore shrinks continuously and the compact is densified. Kingery considered the pressure

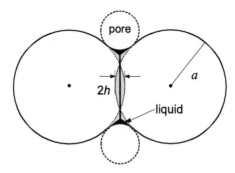

Figure 16.2. Kingery's two-particle model of liquid phase sintering.

---

*A liquid cannot sustain shear stresses and cannot transmit a compressive pressure between particles. The liquid film between the two particles was assumed to behave as a quasi-solid that can transmit the compressive pressure.[103]

difference between the atmosphere and the liquid as the compressive pressure between the particles but, as pointed out in Section 14.2, the surface tension of the liquid also contributes to the compressive pressure. As in the case of the LSW theory, Kingery considered the diffusion of atoms in the liquid and the reaction at the solid/liquid interface to be the controlling mechanisms of material transport. Although the assumptions for reaction control are not justified (see Section 15.2.3), the original equation will be introduced for reference.

Contact flattening controlled by diffusion

The material transport by diffusion through a thin liquid film at the contact area is similar to that in Coble's intermediate stage model.[104] Therefore,

$$J = 4\pi D \Delta C \tag{16.2}$$

and

$$\frac{dV}{dt} = 2\pi a h \frac{dh}{dt}$$

$$\|$$

$$\delta_l J = 4\pi D \delta_l \Delta C \propto 4\pi k_1 D \delta_l C_\infty \frac{\gamma V_m}{hRT} \tag{16.3}$$

Here, δ_l is the liquid film thickness, D the diffusion coefficient of solute atoms in the liquid film, k_1 a constant governed by geometrical parameters, such as pore size and contact area, C_∞ the solubility of the solid with an infinite size in the liquid, and γ the solid/liquid interfacial energy (γ_{sl}). The other parameters are defined in Figure 16.2. Therefore, the shrinkage $\Delta l/l_0$ is expressed as

$$\frac{\Delta l}{l_0} = \frac{h}{a} \approx \frac{1}{3} \frac{\Delta V}{V} = \left(\frac{6k_1 \delta_l D C_\infty \gamma V_m}{RT} \right)^{1/3} a^{-4/3} t^{1/3} \tag{16.4}$$

This equation suggests that the shrinkage is proportional to one-third of the sintering time.

Contact flattening controlled by interface reaction

The rate of material transport controlled by interface reaction was considered to be proportional to a reaction constant and the contact area. Then,

$$\frac{dV}{dt} = 2\pi a h \frac{dh}{dt}$$

$$\|$$

$$k_2 \pi x^2 (a_i - a_{io}) = 2\pi k_2 h a (a_i - a_{io}) \tag{16.5}$$

where k_2 is a proportionality constant, a_i the atom activity at the contact area, and a_{io} the atom activity at the off-contact neck area. Considering that $a_i \approx C$,

$$\frac{\Delta l}{l_0} = \frac{h}{a} \approx \frac{1}{3}\frac{\Delta V}{V} = \left(\frac{2k_1 k_2 C_\infty \gamma V_m}{RT}\right)^{1/2} a^{-1} t^{1/2} \tag{16.6}$$

This equation suggests that the shrinkage is proportional to one-half of the sintering time.

16.1.3 Final Stage: Solid State Sintering

In their sintering model, Cannon and Lenel[6] suggested that, after considerable densification by a solution/reprecipitation process and the formation of grain boundaries, the contribution of solution/reprecipitation became negligible and that final densification occurred by a sintering process similar to solid state sintering. This type of sintering, however, is not believed to be operative in liquid phase sintering because the densification kinetics of liquid phase sintering are much faster than the estimated kinetics of solid state sintering. Kingery suggested no kinetic equation for this questionable stage and, furthermore, accepted the three-stage model of Cannon and Lenel.

16.1.4 Applicability of the Classical Theory

Since the development of the classical model[6] and theory,[7] almost all of the phenomena and kinetics of liquid phase sintering have been explained and analysed in terms of the three-stage model and theory. In particular, the theory of the second stage, contact flattening theory, has been the standard theory of liquid phase sintering for decades, despite doubts raised by some researchers[13,15,16,105,106] about the validity of the basic assumptions. Modified contact flattening theories[9,10] have also been developed, taking grain growth into account to incorporate the pore filling that was observed to be an essential phenomenon occurring during liquid phase sintering.[1,3,11,14] Whether or not grain growth is considered, the contact flattening model is basically a two-particle model, similar to that for the initial stage of solid state sintering.[107] According to this theory, the densification and, hence, the shrinkage of a powder compact is achieved by material transport from the contact area between grains through a thin liquid film to the off-contact neck region, as in solid state sintering. This process results in a continuous reduction in pore size and a continuous change in grain shape that should become more and more anhedral until there is a complete elimination of the pores.

The contact flattening theory contains some inherent problems. One is the continuous reduction in pore size with increasing sintering time. Therefore, in the pore size distribution, the maximum pore size and the frequency of large pores must decrease continuously as the sintering proceeds. Such a change in pore size distribution, however, has not been observed in real sintering. Instead, the frequency of small pores decreased and that of large pores remained until the final densification. Another problem is the continuous change in grain shape during densification. The continuous grain shape change means that a driving force for the shape change is available as long as pores are present in the compact. This is quite unacceptable because most of the grains are immersed in a liquid under a hydrostatic pressure from the beginning of conventional liquid phase sintering. (See Section 3.3 for the equilibrium microstructure of two-phase systems. Also refer to Problem 6.20.) It is also unrealistic to say that the grain shape is very different before and after the final densification. Indeed, the grain shape does not appear to change with densification in real microstructure development.[1] In addition, the theory assumes the presence of a liquid film. This assumption, however, seriously limits the applicability of the theory to real systems. When the dihedral angle is greater than 0° and no liquid film is present in the contact area, which is common for most of the liquid phase sintering systems, the densification should occur very slowly by grain boundary diffusion or volume diffusion and the kinetics should become similar to those of solid state sintering.[108] However, densification in real systems is much faster than predicted by contact flattening,[108] indicating that this type of densification is negligible in liquid phase sintering.

A system that would satisfy the basic assumptions for contact flattening would contain monosize particles and a very small amount of liquid. For such a system, the liquid is localized in the neck region for a wetting angle close to 0° (see Figure 14.5(a)) and grain growth can be neglected. The densification can then be explained using Kingery's two-particle model (Figure 16.2). However, even under this condition densification by contact flattening is limited because of subsequent pore filling.[109]

Figure 16.3 shows the calculated densification curves with liquid volume fraction using the two-particle model for a compact with a face-centred cubic packing of particles.[109] (The constants used in the calculation were typical values in liquid phase sintering: wetting and dihedral angle of 0°, molar volume of 10^{-29} m^3, diffusion coefficient of 10^{-9} m^2/s in the liquid, γ_l of 1 J/m, γ_l/γ_{sl} of 4 and kT of 10^{-20} J.) The small open circles on the densification curves represent the moments when the continuous pore channel along three grain edges is broken and the pores become isolated. The small closed circles represent the critical moment for liquid filling of the pores at tetrahedral sites. The densification curves in Figure 16.3 are characterized by a reduction in densification rate with increasing sintering time and increasing liquid volume fraction f_l. This result is due to the fact that the liquid capillary pressure decreases with increasing liquid meniscus radius during sintering and also with increasing f_l.

The densification curves in Figure 16.3 represent the maximum contribution of contact flattening in a compact with an idealized geometry. In real systems,

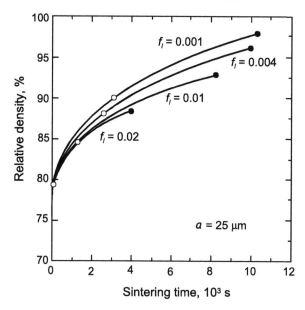

Figure 16.3. Calculated increase in relative density with sintering time for various liquid volume fractions under the assumption of no grain growth.[109] See text for physical constants used in the calculation.

however, such a condition is not satisfied. Contact flattening must be limited to the process that makes the grains attain their equilibrium shape during heating of the compact to a liquid phase sintering temperature and at a very early stage of liquid phase sintering. The attainment of an equilibrium grain shape for a given volume fraction of liquid involves not only contact flattening but also grain growth as recently discussed by Lee and Kang.[5]

The discussion so far suggests that contact flattening is marginal in real liquid phase sintering. It can be operative at the very early stage of liquid phase sintering to contribute to attaining an equilibrium shape of grains. However, once the shape of the grains reaches equilibrium for a given volume fraction of liquid, there is essentially no driving force for further contact flattening and densification. As recently shown by a calculation,[5] the contribution of contact flattening to densification is limited to the very early stage of specific systems with a dihedral angle of 0° and to point contact between particles. Under other conditions densification is predicted to occur by pore filling.

16.2 PORE FILLING MODEL AND THEORY

16.2.1 Development of the Pore Filling Model

The pore filling model of liquid phase sintering was developed via a series of experimental observations.[1,3,11,14–16,110] Kwon and Yoon[11,14] first observed microstructural change during liquid phase sintering in the W-Ni system and suggested a three-stage model:

(i) liquid coagulation,
(ii) liquid redistribution, and
(iii) liquid filling of pores

that describes the behaviour of liquid rather than the classical model. The liquid coagulation in the centre of the sample is a kind of liquid flow to minimize the total liquid/vapour interfacial energy for a solid skeleton of grains in the compact.[111] On the other hand, the liquid redistribution is a process whereby a homogeneous microstructure is obtained with a more or less uniform distribution of pores. The two stages occur in a relatively short sintering time or even before reaching the liquid phase sintering temperature when the compact is slowly heated up.[112] Therefore, the overall sintering kinetics is governed by the third stage — the pore filling stage.

The process of pore filling can be explained by referring to the microstructural evolution observed previously in a model system Mo-Ni.[3] Early experimental observations of pore filling were made in some model systems that contained spherical particles with a melting point lower than that of the matrix particles. Figure 16.4(a) shows a pore formed at the site of a large Ni particle of approximately 100 μm size in a Mo-Ni compact.[3] If a viscous flow of grain–liquid mixture does not occur, as is usual, such a pore is stable for a certain period of sintering time and the grains surrounding the pore grow to match the shape of the pore. The lateral growth of the surrounding grains is demonstrated for a cyclic sintering treatment of a Mo-Ni sample in Figure 16.4(a). The boundaries revealed within individual grains are the layers formed by precipitation of material during cooling and reheating[113] and thus show the shape of the grain after each sintering cycle. It is also clear that the pore did not shrink continuously with sintering time, contrary to the suggestion of Kingery's theory. After a certain period of time, however, the pore is eliminated instantaneously by liquid filling, forming a liquid pocket at its site, as shown in Figure 16.4(b). (The concave solid/liquid interfaces surrounding the liquid pocket in the figure reveal that a large pore had been present at the liquid pocket site and that it was filled with liquid just before the cooling of the sample.) After filling of the pores, the liquid pocket is homogenized by preferential material deposition at the concave solid/liquid interfaces surrounding the liquid pocket. The etch boundaries in grain A towards the liquid pocket in Figure 16.5 show preferential deposition of material leading to microstructural homogenization.[3] Similar pore filling was later observed also in real systems with natural pores, as shown in Figure 16.6 (liquid pockets within circles).[1]

The driving force for pore filling is the difference in liquid pressure.[12] Figure 16.7 shows a schematic diagram of the microstructures at the sample

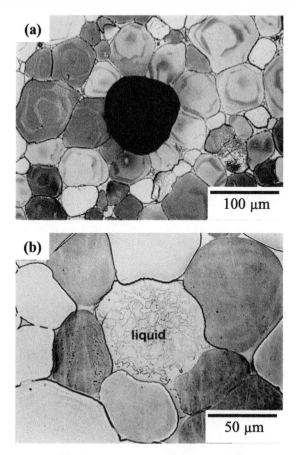

Figure 16.4. Microstructures showing (a) growth pattern of Mo grains around a large isolated pore in a 96Mo-4Ni(wt%) specimen sintered in three cycles (30 + 30 + 30 min) at 1460°C and etched in Murakami solution for about 5 min, and (b) a liquid pocket formed at an isolated pore site by liquid filling in a 96Mo-4Ni specimen sintered at 1460°C for 2 h.[3]

surface and the pore surface during grain growth to explain the driving force.[114] If the gas pressure in the pore is the same as that outside the compact, the liquid meniscus radii at the compact surface and the pore surface are the same because of the hydrostatic pressure of the liquid (Figure 16.7(a)). As long as the pore is stable, the microstructure coarsens in a self-similar manner (see Section 3.3) and the liquid meniscus radius increases linearly with grain growth[115] (Figure 16.7(b)). As this radius becomes equal to the pore radius (in the case of a dihedral angle of 0°) with grain growth, the pore surface is completely wetted, as shown schematically in Figure 16.7(b) (the complete wetting of the pore surface and the critical moment for pore filling). An imbalance of liquid pressures at the compact and pore surfaces then arises with

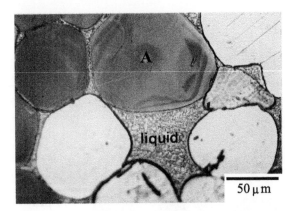

Figure 16.5. Microstructure showing typical growth pattern of a growing grain (A) adjacent to a liquid pocket.[3] 96Mo-4Ni(wt%) specimen sintered in three cycles (60 + 30 + 30 min) at 1460°C and etched in Murakami solution.

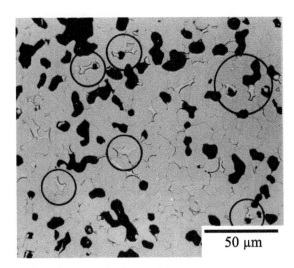

Figure 16.6. Densification of a W(5 μm)-1Ni(4.6 μm)-1Fe(5 μm)(wt%) specimen by liquid filling during liquid phase sintering at 1460°C for 10 min. The circles indicate liquid pockets.[1]

further increase in grain size, because the liquid meniscus radius at the pore surface is limited by the pore size while that at the compact surface is not (Figure 16.7(c)). This imbalance induces a flow of liquid into the pore (pore filling). Since the critical radius for pore filling is proportional to the pore radius, the critical grain size also increases linearly with the pore size.[12] Pore filling occurs therefore in sequence: smaller pores earlier and larger pores later, which

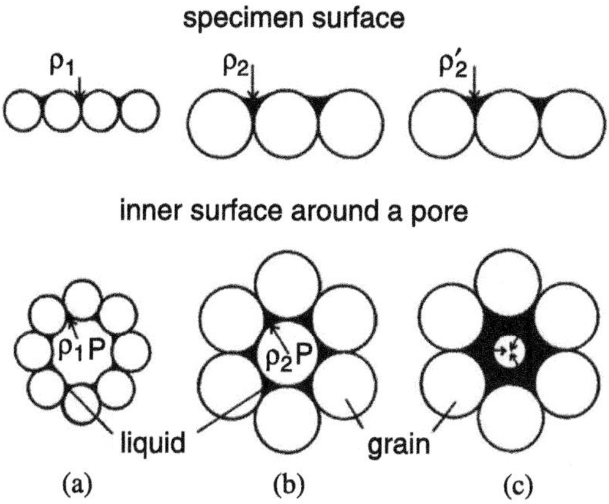

specimen surface

inner surface around a pore

liquid   grain

(a)   (b)   (c)

Figure 16.7. Schematic showing the liquid filling of a pore during grain growth: (a) before pore filling, (b) the critical moment for filling, and (c) liquid flow straight after the critical moment. P is the pore and ρ is the radius of curvature of the liquid meniscus ($\rho_1 < \rho_2$, $\rho_2 < \rho_2'$).[114]

is in agreement with experimental observation.[1] This result means that the compact is densified with grain growth (grain-growth-induced densification).

Based on this earlier work, Kang *et al.*[13] proposed a model of liquid phase sintering (Figure 16.8). Figure 16.8(a) depicts schematically the microstructure of a compact containing pores of various sizes. As long as they are not completely wetted, the pores are stable and the surrounding grains grow around them and since the volume fraction of liquid surrounding the grains does not change with grain growth, there is no driving force for grain shape change. This means that the pores behave like intact third-phase particles within a fully dense solid–liquid two-phase system. Therefore, the radius ratio of grain to liquid meniscus is unchanged during grain growth as in a fully dense system.

However, once the surface of a smaller pore is completely wet as a result of grain growth to a critical size, the liquid spontaneously fills the pore, as illustrated in Figure 16.8(b). With pore filling, the compact density measured by the water-immersion method must increase. In terms of microstructure, however, the result of the pore filling is that the liquid menisci recede at specimen and intact pore surfaces, and thus there is a sudden decrease in liquid pressure. The pressure decrease can also be understood as a reduction of the effective liquid volume surrounding grains remote from the liquid pockets formed. The situation is similar to the suction of liquid from a dense compact by pores, resulting in a substantially lower fraction of liquid for each grain, as in a model experiment.[16]

Because of the liquid pressure decrease, i.e. the capillary pressure increase by pore filling, the shape of the grains tends to become more anhedral during their growth in order to meet the sudden change in liquid pressure (see Section 3.3).

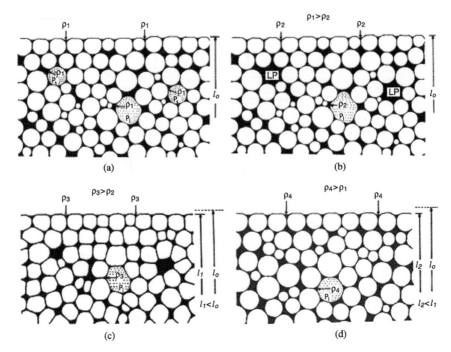

Figure 16.8. Illustration of the process of pore filling and shape accommodation during liquid phase sintering: (a) just before the liquid filling of small pores (P_s) (at a critical condition), (b) directly after the filling of small pores, (c) grain shape accommodation by grain growth and homogenization of microstructure around the liquid pockets formed at the pore sites during prolonged sintering, and (d) just before the liquid filling of a large pore (P_l).[13] For simplicity, in order to show grain shape accommodation during grain growth, the grains are drawn as extremely anhedral. ρ_i is the radius of curvature of the liquid meniscus, l_i the specimen length, and LP the liquid pocket.

Meanwhile, homogenization of the microstructure around the liquid pockets formed proceeds with grain growth (Figure 16.8(c)), as observed in the microstructure shown in Figure 16.5, resulting once more in a homogeneous microstructure containing large stable pores (Figure 16.8(d)). Specimen shrinkage is expected to occur during this microstructural homogenization (Figures 16.8(c) and (d)). In this sense, densification in liquid phase sintering is not directly related to sample shrinkage, unlike solid state sintering where densification and shrinkage have the same meaning. In reality with a pore size distribution, however, the two different phenomena are not expected to occur sequentially but continuously and differently depending on the size of the pores.

The fundamental difference between the pore filling model and the contact flattening model is the material transport mechanism for densification. In the former, the mechanism is liquid flow as a result of grain growth while, in the latter, it is atom-by-atom transport from the contact area to the off-contact area under the capillary pressure of a liquid. As a result, the dependence of densification on grain size in the two models is opposite to each other. For a given pore size distribution, use of coarse powder enhances densification in the pore filling model while it retards densification in the contact flattening model. Recent experimental results[116,117] support the pore filling model consequence that densification is enhanced by use of a coarser powder.

16.2.2 Pore Filling Theory: Grain-Growth-Induced Densification

Based on the pore filling model of liquid phase sintering, Lee and Kang[8] proposed a new theory of liquid phase sintering, namely, the pore filling theory. This section describes the fundamentals of the pore filling theory. To predict the sintering kinetics, the critical condition for the liquid filling of pores must be calculated. In other words, for a given pore size distribution in a liquid phase sintering compact, the evaluation of the radius of liquid meniscus during pore filling and microstructural homogenization is critical for quantitative evaluation of densification. The liquid meniscus radius is determined by the size, shape and packing geometry of grains, the effective liquid volume fraction, the wetting and dihedral angle, and the ratio of liquid/vapour interfacial energy to solid/liquid interfacial energy.[118] For the calculation, closely packed mono-size grains (cubic-close-packing) are assumed.

In order to calculate the liquid meniscus radius during densification, it is critical to estimate the effective volume fraction of liquid, f_l^{eff}, because the liquid meniscus radius depends strongly on f_l^{eff}. The effective volume fraction of liquid in the compact is determined by the initial liquid volume fraction f_l, the volume of pores filled with liquid (liquid pockets) and the homogenization of liquid pockets, as explained in Figure 16.8. When liquid fills a pore, the effective liquid volume in the compact decreases as much as the pore volume. With grain growth, this liquid-filled pore (liquid pocket) is homogenized, and the liquid in the pocket is squeezed out and added to the effective liquid volume. When the microstructural homogenization is completed, the effective liquid volume reaches its initial volume before the pore filling.

Assuming that the microstructural homogenization after pore filling occurs by concentric grain growth towards the liquid pocket centre, the homogenized volume V_{homo}^j of a liquid pocket j can be expressed as

$$V_{homo}^j = -\int_0^t 4\pi r_\tau^2 \left(\frac{dr_\tau}{d\tau}\right) d\tau \qquad (16.7)$$

Here, r_τ is the radius of liquid pocket j being homogenized at time $t = \tau$, and its variation with time is expressed as

$$\frac{dr_\tau}{d\tau} = -\frac{1}{2}\frac{dG}{d\tau} \qquad (16.8)$$

Then, the effective liquid volume fraction f_l^{eff}, relative density ρ and compact shrinkage $(1-l/l_o)$ are expressed, respectively, as

$$f_l^{eff} = \frac{V_l^i - \sum_{j=k+1}^{n} (V_p^j - V_{homo}^j)}{V_s^i + V_l^i - \sum_{j=k+1}^{n} (V_p^j - V_{homo}^j)} \tag{16.9}$$

$$\rho = 1 - \frac{\sum_{j=n+1} V_p^j}{V_s^i + V_l^i + \sum_{j=n+1} V_p^j} \tag{16.10}$$

and

$$1 - \frac{l}{l_o} = 1 - \left(1 - \frac{\sum_{j=k+1}^{n} V_{homo}^j}{l_o^3} - \frac{\sum_{j=1}^{k} V_p^j}{l_o^3}\right)^{1/3} \tag{16.11}$$

Here, V_l^i is the initial volume of liquid (the total volume of liquid in the compact), V_s^i the initial volume of solid, V_p^j the volume of pore j filled with liquid for $j \leq n$ or the volume of unfilled pore j for $j \geq n+1$, l_o the initial dimension of the compact, l the dimension of the compact at time t, and k the maximum size of the completely homogenized liquid-filled pores (liquid pockets).

Since densification and microstructure homogenization are governed by grain growth in the pore filling theory, it is also essential to determine the grain growth kinetics. In their calculation Lee and Kang[8] assumed a diffusion-controlled grain growth, as an example, which follows a cubic law, $G^3 - G_o^3 = Kt$. Here, the proportionality constant K is a function of liquid volume fraction. For the dependence of K on f_l, they used the equation

$$K = K_o \left(\frac{0.05}{f_l^{eff}}\right)^{0.8} \tag{16.12}$$

which was obtained from previous experimental data for the Co-Cu system.[46] Here, K_o is a constant independent of f_l.

Using Eqs (16.7)–(16.12) Lee and Kang[8,119] evaluated the effects of various processing parameters, such as pore size distribution, pore and liquid volume fraction, dihedral and wetting angle, particle size (scale), entrapped gas, etc. Figure 16.9 shows the calculated sintering time to 99.5% relative density of the compacts containing initial pores with a log-normal distribution from one to 30 times larger than the initial grain size for various liquid and pore volume fractions (f_l^i of 2–8% and V_p^i of 5–15%). The grain growth kinetics were assumed to follow a cubic law and K_o/G_o^3 was taken to be $0.5\,\text{s}^{-1}$. Taking a compact containing 5 vol.% liquid and 10 vol% pores as a standard sample, the sintering time is proportional to $\sim (f_l^i)^{-2.9}$ and $(V_p^i)^{2.2}$. If the dependence of grain growth on f_l is negligible, the sintering time for pore filling is

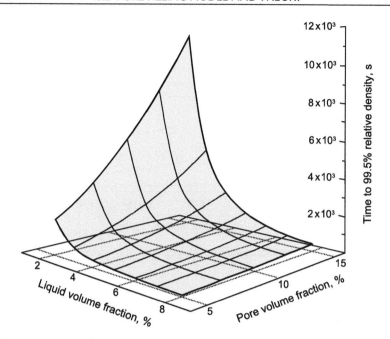

Figure 16.9. Calculated sintering time to 99.5% relative density with various liquid and pore volume fractions for compacts containing initial pores with a log normal distribution from one to 30 times larger than the initial grain size.

proportional to f_l^{-3} because the grain size increases with $t^{1/3}$ (diffusion-controlled growth). Otherwise, the exponent of the liquid volume fraction must be larger than -3 (for example, -2.8), as the calculation shows. On the other hand, as the porosity increases, the sintering time also increases. This effect, however, is less significant than that of the liquid volume fraction. This result arises from the fact that the liquid volume fraction directly affects f_l^{eff} over the whole compact while only a fraction of the pores do so, i.e. the unhomogenized liquid-filled ones.

Lee and Kang[8] also calculated densification curves under various experimental conditions. Figure 16.10 shows the effect of scale on densification. Since the densification in the pore filling theory occurs as a result of grain growth, the time needed for densification is proportional to the time needed for grain growth. Therefore, the exponent in a scaling law for densification is equal to that for grain growth and is 3 in diffusion-controlled grain growth. It should be noted in Figure 16.10 that sample shrinkage occurs even after full densification. This result is related to the fact that the shrinkage occurs continuously with microstructural homogenization after instantaneous liquid filling of pores (see Figure 16.8). The shrinkage curves in Figure 16.10 thus represent the times needed for microstructure homogenization. The dependence of shrinkage on scale is also the same as that of densification.

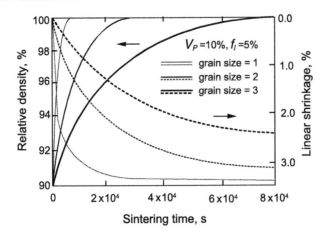

Figure 16.10. Calculated densification and shrinkage curves of compacts with different scales: 1, 2, and 3.[8]

The calculated sintering kinetics are based on the estimation of the radius of the liquid meniscus which is proportional to grain size for a given volume fraction of liquid. The linear relationship between the radius of liquid meniscus and grain size should be satisfied also in systems with faceted grains.[117] For faceted grains, however, it would be difficult to quantitatively evaluate the liquid meniscus radius and describe the sintering kinetics. Nevertheless, the overall behaviour of sintering and the effect of various sintering parameters would be similar to those evaluated for compacts with spherical grains. The difficulties observed in densifying compacts with faceted grains can then be understood as a result of the low growth rate of faceted grains.

16.2.3 Microstructural Development

In the pore filling model[13] and theory,[8] pore filling, i.e. densification occurs as a result of grain growth. Therefore, the grain growth rate directly affects the densification, as shown in Figure 16.11(a) where all other parameters are assumed to be constant except the grain growth constant K_o. As K_o increases, the densification time decreases in proportion. In fact, all of the densification curves in Figure 16.11(a) are reduced to one curve, regardless of K_o, in the relative density versus average grain size diagram in Figure 16.11(b).[119,120]

The microstructure development map in Figure 16.11(b) can be characterized by the slope of the trajectory, $d\rho/dG$. Calculation shows that

$$\frac{d\rho}{dG} \propto \rho^2 \frac{dV_p^{fill}}{dr_p^{fill}} \frac{dr_p^{fill}}{dG} \tag{16.13}$$

(a)

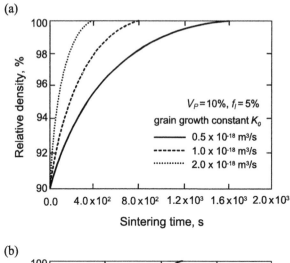

(b)

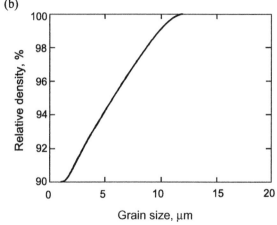

Figure 16.11. Calculated curve of (a) relative density versus sintering time, and (b) relative density versus average grain size for different grain growth constants K_o.[119,120]

where V_p^{fill} is the total volume of liquid-filled pores and r_p^{fill} the maximum size of the liquid-filled pores.[120] The first term, dV_p^{fill}/dr_p^{fill}, on the right-hand side of Eq. (16.13) is determined by the pore size distribution. On the other hand, the second term, dr_p^{fill}/dG, is dependent on pore filling and microstructure homogenization; in general, it decreases at the beginning of sintering, reaches a minimum and then increases.[120] This behaviour is exactly opposite to that of dV_p^{fill}/dr_p^{fill}. The slope of the ρ versus G trajectory (microstructure development) is nearly constant, as in the example in Figure 16.11(b).

Lee and Kang[120] further showed the effects of other processing and sintering parameters, including sintering temperature, initial porosity, average pore size, liquid volume fraction, dihedral and wetting angle, and sintering atmosphere on a relative density–grain size plane. Such understanding of liquid phase sintering in terms of a density–grain size trajectory, i.e. microstructure development, has implications similar to that of the microstructure development in solid state sintering (see Section 11.4).

16.3 ENTRAPPED GASES AND PORE FILLING

When the gas pressure in an isolated pore is different to that outside the compact, pore filling is either retarded with an excess internal pressure or accelerated with an excess external pressure. The former is due to entrapped insoluble gases and, in this case, pore filling is not complete and gas bubbles remain within liquid-filled pores. (Gas bubbles are, in fact, often observed in liquid phase sintered compacts.) Even in compacts with entrapped gases, pore filling can be critical because the possibility of formation of sharp internal notches at the pore surface during solidification of the liquid must be considerably reduced by liquid filling.

Figure 16.12 is a schematic showing the compact surface and the internal surface of a pore containing insoluble gases.[121] Since a hydrostatic pressure is maintained in a liquid, the bulk liquid pressure at the surface is the same as that at the pore surface. Therefore,

$$P_l = P_s - \frac{2\gamma_l}{\rho_s}$$

$$\|$$

$$P_l = \Delta P_p + P_s - \frac{2\gamma_l}{\rho_p} \tag{16.14}$$

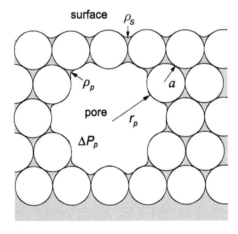

Figure 16.12. Illustration of liquid menisci at the specimen surface and around a pore containing an inert gas of pressure ΔP_p during liquid phase sintering.[121]

where P_l is the liquid pressure, P_s the sintering atmosphere pressure, ΔP_p the difference in gas pressure between the pore and the atmosphere, and ρ_s and ρ_p the radii of liquid menisci at the pore and at the compact surface, respectively. Since ρ_s is linearly proportional to the grain radius a,

$$\rho_s(t) \propto a(t) \tag{16.15}$$

Even under $\Delta P_p \neq 0$, ρ_p is equal to r_p at the critical condition for the wetting of a pore surface. Let $a(\Delta P_p)$ and $a(0)$ be the critical grain sizes under $\Delta P_p \neq 0$ and $\Delta P_p = 0$, respectively. Then,[121]

$$\frac{a(\Delta P_p)}{a(0)} = \frac{\rho_s(\Delta P_p)}{\rho_s(0)} = \frac{\rho_s(\Delta P_p)}{r_p} \tag{16.16}$$

and

$$\frac{a(\Delta P_p)}{a(0)} = \frac{1}{1 - \frac{1}{2}r_p \Delta P_p / \gamma_l} \tag{16.17}$$

Figure 16.13 shows the variation of the calculated critical grain radius with excess internal pressure (Eq. (16.17)) together with experimental data.[122] The positive and negative values in the ordinate, respectively, mean a positive and a negative pressure in the pores relative to atmospheric pressure. When $\Delta P_p > 2\gamma_l/r_p$ in Eq. (16.17) because of either high internal pressure or large pore size, pore filling cannot occur by pressureless sintering and an excess external pressure is needed to fill the pores. On the other hand, $\Delta P_p < 0$ means an excess external pressure which can be provided by applying an external

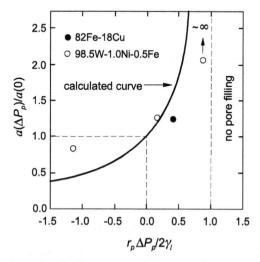

Figure 16.13. Calculated (Eq. (16.17)) and measured critical grain radius ratio, $a(\Delta P_p)/a(0)$, with excess internal gas pressure, ΔP_p.[122]

pressure or by changing the sintering atmosphere after pore isolation. The condition for $\rho_s = \infty$ in Figure 16.13 can be satisfied by immersing the sintered compact within a liquid pool. The experimental data shown in Figure 16.13 fit the predicted curve fairly well, suggesting that theoretical prediction of the gas effect on densification is possible using the pore filling theory. In reality, however, unknown insoluble gases are often produced from the raw powder during sintering[121–124] and retard the densification to a greater extent than predicted. The large discrepancy between the experimental data and the predicted value for $\Delta P_p < 0$ in Figure 16.13 must be due to such insoluble gases. The excess pressure of entrapped gases can be calculated by measuring the size of gas bubbles remaining in liquid pockets or the maximum size of liquid pockets formed after immersing the sintered compact within a liquid pool.[114,121]

Figure 16.14 shows the effects of internal and external excess gas pressures on the microstructure development calculated using the pore filling theory.[120] When applying an excess external pressure of a few atm, the grain size for full densification of a compact being sintered in a fast diffusing gas atmosphere is greatly reduced, indicating the remarkable effect of a small increase in external pressure on densification in liquid phase sintering. On the other hand, when the liquid phase sintering is done in an insoluble inert gas atmosphere of 1 atm, the sintered density is limited, as shown in Figure 16.14. (If pore coalescence occurs, the sintered density decreases, as explained later.) However, the initial densification of a compact containing an inert gas is predicted to be faster than that of a compact without entrapped inert gas. This result is due to a smaller reduction in the effective liquid volume with pore filling in the case of inert gas sintering which results in earlier liquid filling of larger pores.[8,120]

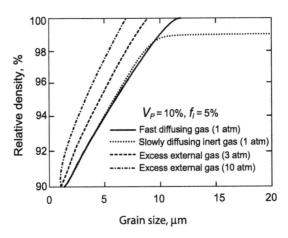

Figure 16.14. Relative density–grain size trajectories for compacts in atmospheres of fast and slowly diffusing gases (1 atm) and in gas pressure sintering (3 and 10 atm).[120] For the calculation, no pore coalescence is assumed.

Isolated pores containing inert gases can coalesce with each other with grain growth and their average size can increase. Oh *et al.*[125] observed that the pore to grain number ratio was constant but the porosity increased with an increased sintering time in an $MgO-CaMgSiO_4$ compact sintered in N_2. As explained by Oh *et al.*,[125] the constant number ratio is due to pore coalescence by grain growth, while porosity increase is due to a reduction in capillary pressure of pores with pore size increase. Therefore, pore coalescence by grain growth in an inert gas atmosphere can reduce the sintered density, although grain growth induces pore filling and densification. To prevent density reduction (dedensification) at final stage sintering in an inert gas, grain growth must be suppressed, as in the case of solid state sintering.[126]

16.4 POWDER COMPACTS AND DENSIFICATION

Unlike solid state sintering, a backstress problem is not involved in liquid phase sintering because of the liquid phase present between solid grains. The densification during liquid phase sintering of a specific system is largely dependent on pore size and distribution.[8] Since the pore filling occurs as a result of grain growth, the sintering time for densification is governed by grain growth kinetics. In the case of diffusion-controlled growth, the sintering time is proportional to the cube of pore size. Therefore, to ensure a fine microstructure after full densification, it is necessary to minimize the generation of large pores, which are due mainly to local densification during heating[2,4] and the melting of liquid-forming particles.[3,11,14] Since local densification is usually due to inhomogeneous mixing and packing of powders,[127,128] homogeneity in these is essential for enhancing the densification kinetics during liquid phase sintering, as in solid state sintering. Intrinsic pores are formed at the sites of liquid-forming particles when they melt, and the melt is sucked out between the solid particles by capillary action (see Section 16.2.1). Therefore, the use of a liquid forming powder of fine size is also necessary to ensure fast densification.

PROBLEMS

6.1. For a two-particle model with a constant volume fraction of liquid, explain the variation of the compressive pressure between the particles with particle size.

6.2. (a) Given that the wetting angle is $0°$, the particle radius a, the two principal radii of liquid meniscus ρ_1 and ρ_2, and the contact angle of liquid ψ (see Figure 14.2), what is the compressive force between two particles of an equal size?

    (b) For a model system of mono-size particles, how does the densification rate $d\rho/dt$ of a compact vary with particle size? Assume no grain growth.

6.3. (a) Calculate the force between a cone-shaped particle and a plate shown in Figure P6.3. For this system, $r_1 = a\{1 - [1 - \cos\alpha/\tan\alpha(1 + \sin\alpha)]\}$ and $r_2 = a/[\tan\alpha(1 + \sin\alpha)]$. Assume that the wetting angle is $0°$.

    (b) What is the dependency of the force calculated in (a) on liquid volume V_l?

    (c) For $\alpha = 30°$ and $\theta = 60°$, what is the force between the two particles?

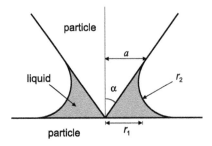

Figure P6.3

6.4. Describe the basic assumptions of the LSW theory and discuss its results and implication.

6.5. Consider two separate particles with a radius of r_1 and r_2, respectively, in a liquid. Draw schematically the solute distributions in the liquid between the particles for diffusion and reaction-controlled growth, respectively. For reaction-controlled growth, are there any differences between the solute distribution in the LSW theory and those in dissolution-controlled growth and precipitation-controlled growth? Explain.

6.6. Consider a liquid phase sintering material containing inert second-phase spheres. Sketch and explain the growth shapes of a grain around a sphere for a dihedral angle between them of 0, 90 and 160°, respectively.

6.7. Consider two liquid phase sintered compacts with the same liquid volume fraction but with different grain sizes. Describe the possible microstructural evolution expected during annealing of the compacts in contact at the liquid phase sintering temperature. Assume that the compacts are fully dense.

6.8. Discuss the similarities and differences between the solution/reprecipitation mechanism in liquid phase sintering and the evaporation/condensation mechanism in solid state sintering.

6.9. Consider a liquid phase sintered compact with a small amount of liquid which forms an interconnected channel along the grain edges (dihedral angle smaller than 60°). Explain a possible method to measure the activation energy of grain growth and discuss the effect of grain boundaries on the grain growth.

6.10. As a mechanism of grain growth in a liquid, the grain coalescence mechanism[129–132] was once proposed in addition to the solution/reprecipitation mechanism. This mechanism seems to suggest that if two grains are in contact with similar crystallographic orientations, they can immediately become one grain. The kinetic equation formed by this mechanism was suggested to be similar to that of the LSW theory, namely, the cubic law of the diffusion-controlled Ostwald ripening. Design an experiment which can judge if the coalescence mechanism is operative in real sintering.

6.11. Consider a two-dimensional crystal with {10} and {11} surface energies of 0.5 and 0.45 J/m^2, respectively. Assuming that the surface energies of the other planes are much higher than these values, delineate an exact equilibrium shape of this crystal.

6.12. The equilibrium shape of an oxide in a melt is reported to be a cube. When you immerse a rectangular cuboid-shaped single crystal of the oxide in the melt for a long time, what will happen? Explain the process in detail in terms of solubility.

6.13. When a WC-Co powder compact is sintered at a liquid phase sintering temperature, often, abnormal grain growth occurs. Explain possible causes of the abnormal grain growth in this alloy. What are possible measures (at least two) to suppress the abnormal grain growth?

6.14. Consider two kinds of powder prepared by different methods but with the same chemical composition, same shape and same average size and distribution. One was severely ball-milled during the preparation and the other was not. Given that the shape of the particles in a liquid is well-faceted, what would be the difference in grain growth mode of the two powders in the liquid? Why?

6.15. The solid/liquid interfacial energy and its anisotropy in a grain can vary with dopant and oxygen partial pressure. Knowing that two liquid phase sintered compacts with the same initial composition and the same initial grain size and distribution showed normal and abnormal grain growth under different atmospheres, respectively, explain possible causes of the difference in grain growth mode between the two compacts.

6.16. Consider two different liquid phase sintering compacts with faceted grains. Starting with the same initial average grain size and initial grain size distribution, what would be the difference in microstructure development between the two compacts if the critical driving force, ΔG^c, for grain growth in each compact is different, small in one and large in the other? Assume that the compacts are fully dense.

6.17. The grain shape of NbC in a Co liquid varies between a well-faceted cube and a sphere depending on annealing temperature and addition of dopants such as B.

    (a) What do you expect as the shape of NbC grains in Co without B at a low temperature?

    (b) Explain the growth mechanisms of spherical and faceted NbC grains in Co. What are the kinetic equations of grain growth in compacts containing the two extreme types of grains?

    (c) Which mechanism do you expect to be operative in the growth of round-edged (partially rounded) cubic NbC grains. Do you think the growth kinetics vary with the fraction of round-edged area? Explain.

6.18. To fabricate a large single crystal from a single crystal seed embedded in a fine powder compact, formation of large grains in the powder compact must be suppressed. Discuss possible measures that can suppress the formation of large grains while enhancing the growth of the seed.

6.19. Explain why densification is, in general, faster in liquid phase sintering than in solid state sintering.

6.20. Consider three different liquid phase sintered compacts with the same chemical composition, same grain size and distribution, and same liquid volume fraction, say 5 vol%. One is fully densified, another is partially densified with a number of pores smaller than the grain size, and another contains only a few pores more than ten times larger than the grain size. Describe and compare the microstructures of these compacts in view of the shapes of the grains within the liquid and in contact with the pores.

6.21. Explain the densification and shrinkage processes in the pore filling model and theory of liquid phase sintering. What are the fundamental differences between this model and Kingery's contact flattening model?

6.22. In liquid phase sintering, pore filling was found to be the essential process of densification. What is the dependence of densification on scale?

6.23. (a) Describe a possible method to estimate the activation energy of densification during liquid phase sintering.
(b) What are the assumptions you made for the estimation?
(c) What is the activation energy of densification in the pore filling theory?

6.24. The pore filling theory of liquid phase sintering predicts that the densification of powder compacts is determined by grain growth. Assuming that the grain growth occurs by diffusion control and all other parameters are invariable:
(a) discuss quantitatively the effect of grain growth rate on densification time, and
(b) plot sintered density versus average grain size trajectories for different grain growth rates with a ratio of 1:2:3.

6.25. According to the pore filling theory, the shrinkage of a compact occurs by a successive accommodation and recovery of grain shape. Explain the process of sample shrinkage in detail and its related driving force.

6.26. Discuss the effects of wetting angle and dihedral angle on densification and grain growth in liquid phase sintering. All other parameters that affect densification and grain growth are assumed to be invariable.

6.27. Explain possible causes for finding mostly large pores in solid state sintered as well as liquid phase sintered compacts.

6.28. Observations suggest that, in general, a solid state sintered compact contains more pores entrapped within grains than a liquid phase sintered compact. Why?

6.29. When you crush a liquid phase sintered compact into coarse powders, as shown in Figure P6.29, make a compact and resinter it, what do you expect as the sintering behaviour with varying liquid volume fraction?

Figure P6.29

6.30. The improvement of densification by an external gas pressure after pore isolation is far more pronounced in liquid phase sintering than in solid state sintering. Application of even a few atm pressure is very effective in liquid phase sintering, in contrast to the case of solid state sintering. Explain why the external pressure effect is different between solid state sintering and liquid phase sintering. What is the implication of your answer?

6.31. Discuss the effect of mixedness of two elemental powders on densification during liquid phase sintering.

REFERENCES

1. Park, J. K., Kang, S.-J. L., Eun, K. Y. and Yoon, D. Y., The microstructural change during liquid phase sintering, *Metall. Trans. A*, **20A**, 837–45, 1989.
2. Huppmann, W. J. and Riegger, H., Modelling of rearrangement processes in liquid phase sintering, *Acta Metall.*, **23**, 965–71, 1975.
3. Kang, S.-J. L., Kaysser, W. A., Petzow, G. and Yoon, D. N., Elimination of pores during liquid phase sintering of Mo-Ni, *Powder Metall.*, **27**, 97–100, 1984.
4. Lee, S.-M., Chaix, J.-M., Martin, C. L., Allibert, C. H. and Kang, S.-J. L., Computer simulation of particle rearrangement in the presence of liquid, *Metals and Materials*, **5**, 197–203, 1999.
5. Lee, S.-M. and Kang, S.-J. L., Evaluation of densification mechanisms of liquid phase sintering, *Z. Metallkd.*, **92**, 669–74, 2001.
6. Cannon, H. S. and Lenel, F. V., Some observations on the mechanism of liquid phase sintering, in *Pulvermetallurgie (Plansee Proceedings 1952)*, F. Benesovsky (ed.), Metallwerk Plansee GmbH, Reutte, 106–22, 1953.
7. Kingery, W. D., Densification during sintering in the presence of a liquid phase. I. Theory, *J. Appl. Phys.*, **30**, 301–306, 1959.
8. Lee, S.-M. and Kang, S.-J. L., Theoretical analysis of liquid phase sintering: pore filling theory, *Acta Mater.*, **46**, 3191–202, 1998.
9. Svoboda, J., Riedel, H. and Gaebel, R., A model for liquid phase sintering, *Acta Mater.*, **44**, 3215–26, 1996.
10. Mortensen, A., Kinetics of densification by solution-reprecipitation, *Acta Mater.*, **45**, 749–58, 1997.
11. Kwon, O. J. and Yoon, D. N., The liquid phase sintering of W-Ni, in *Sintering Processes (Proc. 5th Inter. Conf. on Sintering and Related Phenomena)*, G. C. Kuczynski (ed.), Plenum Press, New York, 208–18, 1980.
12. Park, H.-H., Kwon, O.-J. and Yoon, D. N. The critical grain size for liquid flow into pores during liquid phase sintering, *Metall. Trans. A*, **17A**, 1915–19, 1986.
13. Kang, S.-J. L., Kim, K.-H. and Yoon, D. N., Densification and shrinkage during liquid phase sintering, *J. Am. Ceram. Soc.*, **74**, 425–27, 1991.
14. Kwon, O.-J. and Yoon, D. N., Closure of isolated pores in liquid phase sintering of W-Ni, *Inter. J. Powder Metall. Powder Tech.*, **17**, 127–33, 1981.
15. Kang, S.-J. L., Kaysser, W. A., Petzow, G. and Yoon, D. N., Growth of Mo grains around Al_2O_3 particles during liquid phase sintering, *Acta Metall.*, **33**, 1919–26, 1985.
16. Lee, D. D., Kang, S.-J. L. and Yoon, D. N., A direct observation of the grain shape accommodation during liquid phase sintering, *Scripta Metall.*, **24**, 927–30, 1990.

17. Lifshitz, I. M. and Slyozov, V. V., The kinetics of precipitation from supersaturated solid solutions, *J. Phys. Chem. Solids*, **19**, 35–50, 1961.
18. Wagner, C., Theory of precipitate change by redissolution, *Z. Electrochem.*, **65**, 581–91, 1961.
19. Eremenko, V. N., Naidich, Y. V. and Lavrinenko, I. A., Modelling of capillary forces acting during sintering in the presence of a liquid phase, in *Liquid Phase Sintering*, Consultants Bureau, New York, 55–65, 1970.
20. Heady, R. B. and Cahn, J. W., An analysis of the capillary forces in liquid phase sintering of spherical particles, *Metall. Trans.*, **1**, 185–89, 1970.
21. Trivedi, R. K., Theory of capillarity, in *Lectures on the Theory of Phase Transformations*, Chapter 2., H. I. Aaronson (ed.), AIME, New York, 51–81, 1975.
22. Greenwood, G. W., Particle coarsening, in *The Mechanism of Phase Transformations in Crystalline Solids*, Institute of Metals, London, 103–10, 1969.
23. Burton, W. K., Cabrera, N. and Frank, F. C., The growth of crystals and the equilibrium structure of their surfaces, *Phil. Trans. Roy. Soc. London, A*, **243**, 299–358, 1951.
24. Hirth, J. P. and Pound, G. M., *Condensation and Evaporation*, Pergamon Press, Oxford, 77–148, 1963.
25. Cahn, J. W., On the morphological stability of growing crystals, in *Crystal Growth*, H. S. Peiser (ed.), Pergamon Press, Oxford, 681–90, 1967.
26. Peteves, S. D. and Abbaschian, R., Growth kinetics of solid-liquid Ga interfaces: Part I. Experimental, *Metall. Trans. A*, **22A**, 1259–70, 1991.
27. Kang, S.-J. L. and Han, S.-M., Grain growth in Si_3N_4 based materials, *MRS Bull.*, **20**, 33–37, 1995.
28. Park, Y. J., Hwang, N. M. and Yoon, D. Y., Abnormal growth of faceted (WC) grains in a (Co) liquid matrix, *Metall. Trans. A*, **27A**, 2809–19, 1996.
29. Kang, M.-K., Yoo, Y.-S., Kim, D.-Y. and Hwang, N.-M., Growth of $BaTiO_3$ seed grains by the twin-plane reentrant edge mechanism, *J. Am. Ceram. Soc.*, **83**, 385–90, 2000.
30. Chung, S.-Y. and Kang, S.-J. L., Effect of dislocations on grain growth in $SrTiO_3$, *J. Am. Ceram. Soc.*, **83**, 2828–32, 2000.
31. Chung, S.-Y. and Kang, S.-J. L., Intergranular amorphous films and dislocation-promoted grain growth in $SrTiO_3$, *Acta Mater.*, **51**, 2345–54, 2003.
32. Kang, M.-K., Kim, D.-Y. and Hwang, N. M., Ostwald ripening kinetics of angular grains dispersed in a liquid phase by two-dimensional nucleation and abnormal grain growth, *J. Eu. Ceram. Soc.*, **22**, 603–12, 2002.
33. Rohrer, G. S., Rohrer, C. L. and Mullins, W. W., Coarsening of faceted crystals, *J. Am. Ceram. Soc.*, **85**, 675–82, 2002.
34. Kim, S. S. and Yoon, D. N., Coarsening of Mo grains in the molten Ni-Fe matrix of a low volume fraction, *Acta Metall.*, **33**, 281–86, 1985.
35. Li, C.-Y. and Oriani, R. A., Some considerations on the stability of dispersed systems, in *Oxide Dispersion Strengthening*, G. S. Ansell, T. D. Cooper and F. V. Lenel (eds), Gordon & Breach, New York, 431–64, 1968.
36. Fischmeister, H. and Grimvall, G., Ostwald ripening — a survey, in *Materials Science Research: Sintering and Related Phenomena*, G. C. Kuczynski (ed.), Plenum Press, New York, 119–49, 1973.

37. Lee, D.-D., Kang, S.-J. L. and Yoon, D. N., Mechanism of grain growth and α–β' transformation during liquid phase sintering of β'-Sialon, *J. Am. Ceram. Soc.*, **71**, 803–806, 1988.

38. Han, S.-M. and Kang, S.-J. L., Comment on kinetics of β-Si$_3$N$_4$ grain growth in Si$_3$N$_4$ ceramics sintered under high nitrogen pressure, *J. Am. Ceram. Soc.*, **76**, 3178–79, 1993.

39. Warren, R. and Waldron, M. B., Microstructural development during the liquid phase sintering of cemented carbides, *Powder Metall.*, **15**, 166–201, 1972.

40. Sarian, S. and Weart, H. W., Kinetics of coarsening of spherical particles in a liquid matrix, *J. Appl. Phys.*, **37**, 1675–81, 1966.

41. Ardell, A. J., The effect of volume fraction on particle coarsening: theoretical considerations, *Acta Metall.*, **20**, 61–71, 1972.

42. Kang, T.-K. and Yoon, D. N., Coarsening of tungsten grains in liquid nickel-tungsten matrix, *Metall. Trans. A*, **9A**, 433–38, 1978.

43. Brailsford, A. D. and Wynblatt, P., The dependence of Ostwald ripening kinetics on particle volume fraction, *Acta Metall.*, **27**, 489–97, 1979.

44. Davies, C. K. L., Nash, P. and Stevens, R. N., The effect of volume fraction of precipitate on Ostwald ripening, *Acta Metall.*, **28**, 179–89, 1980.

45. Kang, C. H. and Yoon, D. N., Coarsening of cobalt grains dispersed in liquid copper matrix, *Metall. Trans. A*, **12A**, 61–69, 1981.

46. Kang, S. S. and Yoon, D. N., Kinetics of grain coarsening during sintering of Co-Cu and Fe-Cu alloys with low liquid contents, *Metall. Trans. A*, **13A**, 1405–11, 1982.

47. Hardy, S. C. and Voorhees, P. W., Ostwald ripening in a system with a high volume fraction of coarsening phase, *Metall. Trans. A*, **19A**, 2713–21, 1988.

48. DeHoff, R. T., A geometrically general theory of diffusion controlled coarsening, *Acta Metall. Mater.*, **39**, 2349–60, 1991.

49. Fang, Z. and Patterson, B. R., Experimental investigation of particle size distribution influence on diffusion controlled coarsening, *Acta Metall. Mater.*, **41**, 2017–24, 1993.

50. Yu, J. H., Kim, T. H. and Lee, J. S., Particle growth during liquid phase sintering of nanocomposite W-Cu powder, *Nanostr. Mater.*, **9**, 229–32, 1997.

51. German, R. M. and Olevsky, E. A., Modeling grain growth dependence on the liquid content in liquid phase sintered materials, *Metall. Mater. Trans. A*, **29A**, 3057–66, 1998.

52. Fullman, R. L., Measurement of particle sizes in opaque bodies, *Trans AIME*, **197**, 447–52, 1953.

53. Underwood, E. E., *Quantitative Stereology*, Addison-Wesley, Reading, Mass., 1970.

54. Kang, S. S., Ahn, S. T. and Yoon, D. N., Determination of spherical grain size from the average area of interaction in Ostwald ripening, *Metallography*, **14**, 267–70, 1981.

55. Vander Voort, G. F., *Metallography, Principles and Practice*, ASM International, Materials Park, OH, 1999.

56. Gibbs, J. W., *The Scientific Papers of J. Willard Gibbs, PhD, LLD, Vol. 1 Thermodynamics*, Dover Publ., New York, 314–31, 1961.

57. Herring, C., Some theorems on the free energies of crystal surfaces, *Phys. Review*, **82**, 87–93, 1951. (G. Wulff, Zur Frage der Geschwindigkeit des Wachstums und der Auflösung der Krystallflächen, *Z. Kristallogr.*, **34**, 449–530, 1901.)

58. Semenchenko, V. K., *Surface Phenomena in Metals and Alloys*, Chapter 9. Surface phenomena in solids, Addison-Wesley, Reading, Mass., 272–302, 1962.

59. Murr, L. E., *Interfacial Phenomena in Metals and Alloys*, Chapter 1. Thermodynamics of solid interfaces, Addison-Wesley, Reading, Mass., 1–30, 1975.

60. Mullins, W. W., Solid state morphologies governed by capillarity, in *Metal Surfaces: Structure, Energetics and Kinetics*, ASM, Metals Park, Ohio, 17–66, 1963.

61. Miller, W. A. and Chadwick, G. A., The equilibrium shapes of small liquid droplets in solid–liquid phase mixtures: metallic h.c.p. and metalloid systems, *Proc. Roy. Soc. A*, **312**, 257–76, 1969.

62. Park, S.-Y., Choi, K., Kang, S.-J. L. and Yoon, D. N., Shape of $MgAl_2O_4$ grains in a CaMgSiAlO glass matrix, *J. Am. Ceram. Soc.*, **75**, 216–19, 1992.

63. Choi, J. H., Kim, D.-Y., Hockey, B. J., Wiederhorn, S. M., Handwerker, C. A., Blendell, J. E., Carter, W. C. and Roosen, A. R., Equilibrium shape of internal cavities in sapphire, *J. Am. Ceram. Soc.*, **80**, 62–68, 1997.

64. Kitayama, M. and Glaeser, A. M., The Wulff shape of alumina: III, Undoped alumina, *J. Am. Ceram. Soc.*, **85**, 611–22, 2002.

65. Choi, J.-H., Kim, D.-Y., Hockey, B. J., Wiederhorn, S. M., Blendell, J. E. and Handwerker, C. A., Equilibrium shape of internal cavities in ruby and the effect of surface energy anisotropy on the equilibrium shape, *J. Am. Ceram. Soc.*, **85**, 1841–44, 2002.

66. Adams, B. L., Wright, S. I. and Kunze, K., Orientation imaging: the emergence of a new microscopy, *Metall. Trans. A*, **24**, 819–31, 1993.

67. Saylor, D. M. and Rohrer, G. R., Determining crystal habits from observations of planar sections, *J. Am. Ceram. Soc.*, **85**, 2799–804, 2002.

68. Howe, J. M., *Interfaces in Materials: Atomic Structure, Thermodynamics and Kinetics of Solid-Vapour, Solid-Liquid and Solid-Solid Interfaces*, John Wiley & Sons, New York, 75–86, 1997.

69. Chung, S.-Y., Yoon, D. Y. and Kang, S.-J. L., Effects of donor concentration and oxygen partial pressure on interface morphology and grain growth behaviour in $SrTiO_3$, *Acta Mater.*, **50**, 3361–71, 2002.

70. Warren, R., Microstructure development during the liquid phase sintering of two-phase alloys, with special reference to the NbC/Co system, *J. Mater. Sci.*, **3**, 471–85, 1968.

71. Moon, H., Kim, B.-K. and Kang, S.-J. L., Growth mechanism of round-edged NbC grains in Co liquid, *Acta Mater.*, **49**, 1293–99, 2001.

72. Han, J.-H., Chung, Y.-K., Kim, D.-Y., Cho S.-H. and Yoon, D. N., Temperature dependence of the shape of ZnO grains in a liquid matrix, *Acta Metall.*, **37**, 2705–708, 1989.

73. Frank, F. C., On the kinematic theory of crystal growth and dissolution processes, in *Growth and Perfection of Crystals*, R. H. Doremus, B. W. Roberts and D. Turnbull (eds), Wiley, New York, 411–19, 1958.

74. Frank, F. C., On the kinematic theory of crystal growth and dissolution processes, II, *Z. Phys. Chem. Neue Folge*, **77**, 84–92, 1972.

75. Schreiner, M., Schmitt, Th., Lassner, E. and Lux, B., On the origins of discontinuous grain growth during liquid phase sintering of WC-Co cemented carbides, *Powder Metall. Inter.*, **16**, 180–83, 1984.

76. Hong, B. S., Kang, S.-J. L. and Brook, R. J., The effect of powder purity and sintering temperature on the microstructure of sintered Al_2O_3, unpublished work, 1988.

77. Bae, S. I. and Baik, S., Determination of critical concentrations of silica and/or calcia for abnormal grain growth in alumina, *J. Am. Ceram. Soc.*, **76**, 1065–67, 1993.

78. Kwon, S.-K., Hong, S.-H., Kim, D.-Y. and Hwang, N. M., Coarsening behavior of tricalcium silicate (C_3S) and dicalcium silicate (C_2S) grains dispersed in a clinker melt, *J. Am. Ceram. Soc.*, **83**, 1247–52, 2000.

79. Jang, C.-W., Kim, J. S. and Kang, S.-J. L., Effect of sintering atmosphere on grain shape and grain growth in liquid phase sintered silicon carbide, *J. Am. Ceram. Soc.*, **85**, 1281–84, 2002.

80. Jung, Y.-I., Choi, S.-Y. and Kang, S.-J. L., Grain growth behaviour during stepwise sintering of barium titanate in hydrogen gas and air, *J. Am. Ceram. Soc.*, **86**, 2228–30, 2003.

81. Peteves, S. D. and Abbaschian, R., Growth kinetics of solid–liquid Ga interfaces: Part II. Theoretical, *Metall. Trans. A*, **22A**, 1271–86, 1991.

82. Yoon, D. N. and Huppmann, W. J., Grain growth and densification during liquid phase sintering of W-Ni, *Acta Metall.*, **27**, 693–98, 1979.

83. Cahn, J. W., Hillig, W. B. and Sears, G. W., The molecular mechanism of solidification, *Acta Metall.*, **12**, 1421–39, 1964.

84. Park, C. W. and Yoon, D. Y., Abnormal grain growth in alumina with anorthite liquid and the effect of MgO addition, *J. Am. Ceram. Soc.*, **85**, 1585–93, 2002.

85. Unpublished work by a group of students in an undergraduate laboratory course at the Korea Advanced Institute of Science and Technology, 2003.

86. Kim, S. M., Ko, J. Y. and Yoon, D. Y., Coarsening of cubic TiC grains with round edges in a liquid Ni-rich matrix, presented at the Annual Fall Meeting of the Korean Ceramic Society, Paejae University, Daejeon, Korea, Oct. 17–18, 2003.

87. van Beijeren, H., Exactly solvable model for the roughening transition of a crystal surface, *Phys. Rev. Lett.*, **38**, 993–96, 1977.

88. Wolf, P. E., Gallet, F., Balibar, S., Rolley, E. and Nozières, P., Crystal growth and crystal curvature near roughening transitions in hcp ^{4}He, *J. Physique*, **46**, 1987–2007, 1985.

89. van Beijeren, H. and Nolden, I., The roughening transition, in *Structure and Dynamics of Surfaces II, Phenomena, Models, and Methods*, W. Schommers and P. von Blankenhagen (eds), Springer-Verlag, Berlin, 259–300, 1987.

90. Mitomo, M. and Uenosono, S., Microstructural development during gas-pressure sintering of α-silicon nitride, *J. Am. Ceram. Soc.*, **75**, 103–108, 1992.

91. Hirosaki, N., Akimune, Y. and Mitomo, M., Effect of grain growth of β-silicon nitride on strength, Weibull modulus, and fracture toughness, *J. Am. Ceram. Soc.*, **76**, 1892–94, 1993.

92. Wynblatt, P. and Gjostein, N. A., Particle growth in model supported metal catalysis—I. Theory, *Acta Metall.*, **24**, 1165–74, 1976.

93. Cho, Y. K., Interface roughening transition and grain growth in BaTiO$_3$ and NbC-Co, PhD thesis, KAIST, Daejeon, Korea, 2003.

94. Seabauch, M. W., Kerscht, I. H. and Messing, G. L., Texture development by templated grain growth in liquid phase sintered α-alumina, *J. Am. Ceram. Soc.*, **80**, 1181–88, 1997.

95. Hong, S.-H., Trolier-McKinstry, S. and Messing, G. L., Dielectric and electromechanical properties of textured niobium-doped bismuth titanate ceramics, *J. Am. Ceram. Soc.*, **83**, 113–18, 2000.

96. Fukuchi, E., Kimura, T., Tani, T., Takeuchi, T. and Saito, Y., Effect of potassium concentration on the grain orientation in bismuth sodium potassium titanate, *J. Am. Ceram. Soc.*, **85**, 1461–66, 2002.

97. Khan, A., Meschke, F. A., Li, T., Scotch, A. M., Chan, H. M. and Harmer, M. P., Growth of Pb(Mg$_{1/3}$Nb$_{2/3}$)O$_3$-35 mol% PbTiO$_3$ single crystals from {111} substrates by seeded polycrystal conversion, *J. Am. Ceram. Soc.*, **82**, 2958–62, 1999.

98. Lee, H.-Y., Kim, J.-S. and Kim, D.-Y., Fabrication of BaTiO$_3$ single crystals using secondary abnormal grain growth, *J. Eu. Ceram. Soc.*, **20**, 1595–97, 2000.

99. Fisher, J. G., Kim, M.-S., Lee, H. Y. and Kang, S.-J. L., Effect of Li$_2$O and PbO additions on abnormal grain growth in the Pb(Mg$_{1/3}$Nb$_{2/3}$)O$_3$-35 mol% PbTiO$_3$ system, *J. Am. Ceram. Soc.*, **87**, 937–42, 2004.

100. Lee, H. Y., Solid-state single crystal growth (SSCG) method: a cost-effective way of growing piezoelectric single crystals, in *Piezoelectric Single Crystals and Their Application*, S. Trolier-Mckinstry, L. E. Cross and Y. Yamashita (eds), 160–77, 2004.

101. Lee, S.-M., Chaix, J. M. and Martin, C. L., Computer simulation of particle rearrangement in liquid phase sintering: effect of starting microstructure, in *Sintering Science and Technology*, R. M. German, G. L. Messing and R. G. Cornwall (eds), Penn State University, University Park, 399–404, 2000.

102. Riniger, E. and Raj, R., Packing and sintering of two-dimensional structures made from bimodal particle size distributions, *J. Am. Ceram. Soc.*, **70**, 843–49, 1987.

103. Tu, K.-N., Mayer, J. W. and Feldman, L. C., *Electronic Thin Film Science for Electrical Engineers and Materials Scientists*, Macmillan Publ. Co., New York, 246–80, 1992.

104. Coble, R. L., Sintering of crystalline solids. I. Intermediate and final state diffusion models, *J. Appl. Phys.*, **32**, 789–92, 1961.

105. Kaysser, W. A. and Petzow, G., Ostwald ripening and shrinkage during liquid phase sintering, *Z. Metallkd.*, **76**, 687–92, 1985.

106. Kaysser, W. A., Zivkovic, M. and Petzow, G., Shape accommodation during grain growth in the presence of a liquid phase, *J. Mater. Sci.*, **20**, 578–84, 1985.

107. Kingery, W. D. and Berg, M., Study of the initial stages of sintering solids by viscous flow, evaporation-condensation and self-diffusion, *J. Appl. Phys.*, **26**, 1205–12, 1955.

108. Gessinger, G. H., Fischmeister, H. F. and Lukas, H. L., A model for second-stage liquid phase sintering with a partially wetting liquid, *Acta Metall.*, **21**, 715–24, 1973.

109. Kim, K.-H. and Kang, S.-J. L., Densification of spherical powder compacts containing limited volume of liquid, *Proc. 1993 Powder Metall. World Congress*,

Y. Bando and K. Kosuge (eds), Jap. Soc. Powder and Powder Metall., Kyoto, 357–60, 1993.

110. Kang, S.-J. L. and Azou, P., Trapping of pores and liquid pockets during liquid phase sintering, *Powder Metall.*, **28**, 90–92, 1985.

111. Kim, Y. S., Park, J. K. and Yoon, D. N., Liquid flow into the interior of W-Ni-Fe compacts during liquid phase sintering, *Inter. J. Powder Metall. Powder Tech.*, **20**, 29–37, 1985.

112. Yoo, Y.-S., Kim, J.-J. and Kim, D.-Y., Effect of heating rate on the microstructural evolution during sintering of $BaTiO_3$ ceramics, *J. Am. Ceram. Soc.*, **70**, C322–24, 1987.

113. Kim, S. S. and Yoon, D. N., Coarsening behaviour of Mo grains dispersed in liquid matrix, *Acta Metall.*, **31**, 1151–57, 1983.

114. Kang, S.-J. L., Greil, P., Mitomo, M. and Moon, J.-H., Elimination of large pores during gas-pressure sintering of β-Sialon, *J. Am. Ceram. Soc.*, **72**, 1166–69, 1989.

115. Park, H.-H., Cho, S.-J. and Yoon, D. N., Pore filling process in liquid phase sintering, *Metall. Trans. A*, **15A**, 1075–80, 1984.

116. Baung, J.-C., Choi, Y.-G., Kang, E.-S., Baek, Y.-K., Jung, S.-W. and Kang, S.-J. L., Effects of sintering atmosphere and Ni content on the liquid phase sintering of TiB_2-Ni, *J. Kor. Ceram. Soc.*, **38**, 207–11, 2001.

117. Kim, Y.-P., Jung, S.-W., Kim, B.-K. and Kang, S.-J. L., Enhanced densification of liquid phase sintered WC-Co by use of coarse WC powder: experimental support for the pore filling theory, unpublished work (2003), to be published.

118. Park, H. H., Kang, S.-J. L. and Yoon, D. N., An analysis of surface menisci in a mixture of liquid and deformable grains, *Metall. Trans. A*, **17A**, 325–30, 1986.

119. Kang, S.-J. L. and Lee, S. M., Liquid phase sintering: grain-growth induced densification, in *Sintering Science and Technology*, R. M. German, G. L. Messing and R. G. Cornwall (eds), Penn State University, University Park, 239–46, 2000.

120. Lee, S.-M. and Kang, S.-J. L., Microstructure development during liquid phase sintering, *Z. Metallkd.*, to be published.

121. Cho, S.-J., Kang, S.-J. L. and Yoon, D. N., Effect of entrapped inert gas on pore filling during liquid phase sintering, *Metall. Trans. A*, **17A**, 2175–82, 1986.

122. Kang, S.-J. L. and Yoon, D. N., Morphological changes of pores and grains during liquid phase sintering, in *Horizons of Powder Metallurgy (Proc. 1986 Int. Conf. and Exhib.)*, W. A. Kaysser and W. J. Huppman (eds), Verlag Schmid GmbH, Freiburg, 1214–18, 1986.

123. German, R. M. and Churn, K. S., Sintering atmosphere effects on the ductility of W-Ni-Fe heavy metals, *Metall. Trans. A*, **15A**, 747–54, 1984.

124. Kang, S.-J. L., Hong, B. S., Cho, Y. K., Hwang, N. M. and Yoon, D. N., Residual porosities in liquid phase sintered W-Ni-Fe, in *Sintering '85*, G. C. Kuczynski, D. P. Uskokovic, H. Palmour III, and M. M. Ristic (eds), Plenum Press, New York, 173–78, 1985.

125. Oh, U.-C., Chung, Y.-S., Kim, D.-Y. and Yoon, D. N., Effect of grain growth on pore coalescence during the liquid phase sintering of $MgO-CaMgSiO_4$ systems, *J. Am. Ceram. Soc.*, **71**, 854–57, 1988.

126. Yoon, K.-J. and Kang, S.-J. L., Densification of ceramics containing entrapped gases, *J. Eu. Ceram. Soc.*, **5**, 135–39, 1989.

127. Huppmann, W. J. and Bauer, W., Characterization of the degree of mixing in liquid phase sintering experiments, *Powder Metall.*, **18**, 249–58, 1975.
128. Huppmann, W. J., Riegger, H., Kaysser, W. A., Smolej, V. and Pejovnik, S., The elementary mechanisms of liquid phase sintering, I rearrangement, *Z. Metallkd.*, **70**, 707–13, 1979.
129. Parikh, N. M. and Humenik, Jr., M., Cermets: II. Wettability and microstructural studies in liquid phase sintering, *J. Am. Ceram. Soc.*, **40**, 315–20, 1957.
130. Courtney, T. H. and Lee, J. K., An analysis for estimating the probability of particle coalescence in liquid phase sintered systems, *Metall. Trans A*, **11A**, 943–47, 1980.
131. Takajo, S., Kaysser, W. A. and Petzow, G., Analysis of particle growth by coalescence during liquid phase sintering, *Acta Metall.*, **32**, 107–13, 1984.
132. Kaysser, W. A., Takajo, S. and Petzow, G., Particle growth by coalescence during liquid phase sintering of Fe-Cu, *Acta Metall.*, **32**, 115–22, 1984.

INDEX

Printed and bound by CPI Group (UK) Ltd, Croydon, CR0 4YY

03/10/2024

01040430-0011